Progress in Mathematics
Vol. 16

Edited by
J. Coates and
S. Helgason

Birkhäuser
Boston • Basel • Stuttgart

Phillip A. Griffiths
John W. Morgan

Rational Homotopy Theory and Differential Forms

1981

Birkhäuser
Boston • Basel • Stuttgart

Authors:

Phillip A. Griffiths*
Department of Mathematics
Harvard University
Cambridge, Massachusetts 02139

John W. Morgan**
Department of Mathematics
Columbia University
New York, New York 10027

* Work partially supported by NSF Grant MCS7707782 and the Guggenheim Foundation
**Work partially supported by NSF Grant MCS7904715

Library of Congress Cataloging in Publication Data

Griffiths, Phillip, 1938-
Rational homotopy theory and differential forms.

(Progress in mathematics ; 16)
"Originated as a set of informal notes from a summer course taught by the present authors, together with Eric Friedlander, at the Istituto Matematico 'Ulisse Dini' in Florence during the summer of 1972"--Introd.
Bibliography: p.
Includes Index.
1. Homotopy theory. 2. Differential Forms.
I. Morgan, John W., 1946- . II. Title.
III. Series: Progress in mathematics (Cambridge, Mass.) ; 16.
QA612.7.G74 514'.24 81-12219
ISBN 3-7643-3041-4 AACR2

CIP-Kurztitelaufnahme der Deutschen Bibliothek

Griffiths, Phillip A.:
Rational homotopy theory and differential forms/
Phillip A. Griffiths and John W. Morgan.-
Boston ; Basel ; Stuttgart : Birkhäuser, 1981.
(Progress in mathematics ; Vol. 16)
ISBN 3-7643-3041-4
NE: Morgan, John W.:; GT

ISBN 3-7643-3041-4
Printed in USA

Table of Contents

This monograph originated as a set of informal notes from a summer course taught by the present authors, together with Eric Friedlander, at the Istituto Matematico "Ulisse Dini" in Florence during the summer of 1972. Even though more formal expositions of Sullivan's theory have since appeared, including the major original source[(1)], there has been a steady continuing demand for the old Florence notes. Moreover, one of us (J.M.) has become involved in the subject again through a series of lectures given at the University of Utah in January, 1980, together with joint work in progress with James Carlson and Herb Clemens on a new type of application of the theory to algebraic geometry. Since the Florence notes represented an approach and point of view that does not appear in the literature, we decided to publish the present revised and corrected version.

The material in this monograph is outlined in the table of contents and is informally discussed in the introduction below. Here we should like to observe that the text roughly divides into two parts. The first seven chapters essentially constitute an introductory cause in algebraic topology with emphasis on homotopy theory. The main prerequisite is some familiarity with simplicial homology, covering spaces, and CW complexes.

Chapters VIII-XIV cover the main topic of differential forms and homotopy theory, with emphasis on the homotopy-theoretic and functorial properties of differential graded algebras and minimal models, a topic that does not appear

[(1)] D. Sullivan, "Infinitesimal Calculations in Topology" Publ. I.H.E.S. vol. 47 (1978), 269-331.

explicitly in detail in the literature. An extensive set of exercises, frequently with copious hints, forms an essential complement to the material in the text.

We should like to make several acknowledgements to colleagues whose help and advice has been invaluable. The first and foremost is to Dennis Sullivan. It was he who introduced us to the idea of relating homotopy theory and differential forms, and who explained to us his theory around which these notes are built.

The second is to Francesco Gherardelli who organized the original summer course, and to the Istituto Matematico "Ulisse Dini" and the city of Florence, which together provided excellent mathematical and cultural conditions for the initial preparation of the notes. While in Florence we benefited from conversations with Ngo Van Que, Jim Carlson, and Mark Green. Finally, Moishe Breiner prepared a beautifully handwritten set of notes that constituted the original version of this monograph.

We should also like to thank the University of Utah and above mentioned coworkers of J.M. for providing support and motivation leading to the revision of the Florence notes.

Finally we should like to point out two predecessors of the present theory. The first is Whitney's book "Geometric Integration Theory", Princeton Press (1957). As explained to us by Sullivan, this book contains the genesis of the use of differential forms to solve the commutative cochain problem and thus get the homotopy type of the space. The main thing lacking at the time Whitney wrote the book was the $\mathbb{Q}$-structure. Secondly the relationship between differential forms and homotopy theory was anticipated by K.T. Chen (cf. his recent survey paper: Iterated Path Integrals, Bulletin of the American Mathematical Society, Volume 83, September 1977, pp. 831-879). Many of the results we find from a general viewpoint were established, frequently in stronger form, by him using the method of iterated integrals.

0. Introduction

The overall purpose of this course is to relate the C^∞-differential forms on a manifold to algebraic-topological invariants. A model of results along these lines is deRham's theorem, which says that the cohomology of the differential graded algebra of C^∞ forms is isomorphic to the singular cohomology with $\mathbb{R}$-coefficients, i.e.

$$H^*_{DR}(M) \cong H^*(M,\mathbb{R}) \quad (C^\infty \text{ deRham theorem}).$$

The main theorem of this course will be that from the differential graded algebra of C^∞ forms it is possible to calculate <u>all</u> of the real algebraic-topological invariants of the manifold. More precisely, we shall be able to use the forms to obtain the (Postnikov tower) $\otimes$ $\mathbb{R}$ of the manifold.

In the first 7 lectures of the course we shall discuss the standard terminology, objects, and theorems of elementary homotopy theory, culminating in the description of the Postnikov tower of a space. We then define the <u>localization</u> of a CW complex at $\mathbb{Q}$; this allows us to take a CW complex and replace it by one in which all torsion and divisibility phenomena have been removed (allowing one to focus on the $\mathbb{Q}$-information in the original space). When we compare the Postnikov tower of the original space with that of its localization, we see that all the relevant information (homotopy and homology groups, k-invariants) has been tensored with $\mathbb{Q}$.

Once we have established these basic facts, we turn to the main theorem as shown to us by Sullivan. First we define the rational P.L. forms on a simplicial complex. By integration, these forms give <u>$\mathbb{Q}$-valued</u> cochains. In their own right they form a differential graded algebra/$\mathbb{Q}$ whose cohomology is proved to be isomorphic to that of the space:

$$H^*_{P.L.}(K) \cong H^*(K,\mathbb{Q}) \quad \text{(P.L. deRham theorem)}.$$

There are two very important points here. The first is that we are working over $\mathbb{Q}$ rather than $\mathbb{R}$, as we would be forced to do with C^∞ forms. The second is that the P.L. forms are a differential, graded-commutative algebra--the simplicial or singular cochains over $\mathbb{Q}$ are not commutative. Thus the P.L. forms have a good property of ordinary cochains (they are defined over $\mathbb{Q}$) and a good property of C^∞ forms (they are graded-commutative). Both these properties are essential.

Next we turn to the homotopy of differential graded algebras (D.G.A.'s). Given one such, $\mathfrak{a}$, we show how to extract a <u>minimal model</u> for it. This is a D.G.A., $\mathfrak{M}_{\mathfrak{a}}$, which satisfies some internal condition, together with a map of D.G.A.'s:

$$\rho_{\mathfrak{a}}: \mathfrak{M}_{\mathfrak{a}} \longrightarrow \mathfrak{a}$$

which induces an isomorphism on cohomology.

In the case that $H^1(\mathfrak{a}) = 0$ the internal properties which $\mathfrak{M}_{\mathfrak{a}}$ is required to satisfy are:

(i) it is free as a graded-commutative algebra with generators in degrees ≥ 2 only;

(ii) for all $x \in \mathfrak{M}_{\mathfrak{a}}$, dx is decomposable.

It turns out that these properties characterize $\mathfrak{M}_{\mathfrak{a}}$ up to isomorphism.

We shall show, in addition, that when $\mathfrak{a}$ is the algebra of p.l. forms on a simply connected, simplicial complex X, then $\mathfrak{M}_{\mathfrak{a}}$ is dual to the rational Postnikov tower of X. The duality between minimal models defined over $\mathbb{Q}$ and rational Postnikov towers is described in Section 11. Schematically we have a "commutative diagram":

$$\begin{array}{ccc} \left\{\begin{matrix}\text{simplicial}\\ \text{complexes}\end{matrix}\right\} & \xrightarrow{\quad\text{P.L. Forms}\quad} & \{\text{D.G.A.}/\mathbb{Q}\} \\ \Big\downarrow \text{localization at } \mathbb{Q} & & \Big\downarrow \\ \{\mathbb{Q}\text{-spaces}\} & & \text{minimal model} \\ \Big\downarrow \text{Postnikov tower} & & \Big\downarrow \\ \{\text{Postnikov towers}\} & \longrightarrow & \{\text{Minimal D.G.A.}/\mathbb{Q}\} \end{array}$$

$$\begin{array}{ccc} K & \longleftrightarrow & A^*_{P.L.}(K) \\ \downarrow & & \downarrow \\ K_{(0)} & & \mathfrak{M}(A^*_{P.L.}(K)) \\ \downarrow & & \parallel \\ \left\{\begin{matrix} \vdots \\ \downarrow \\ K_{(0)3} \\ \downarrow \\ K_{(0)2} \end{matrix}\right\} & \longleftrightarrow & \left\{\begin{matrix} \vdots \\ \cup \\ \mathfrak{M}(3) \\ \cup \\ \mathfrak{M}(2) \end{matrix}\right\} \end{array}$$

Given a C^∞ manifold M, we may smoothly triangulate M and have both the C^∞ and P.L. forms. These are both included in the D.G.A. $A^*_{p.C^\infty}(M)$ of "piecewise C^∞ forms" and the inclusions

$$\begin{array}{ccc} A^*_{C^\infty}(M) & \searrow & \\ & & A^*_{p.C^\infty}(M) \\ A^*_{P.L.}(M) \otimes_{\mathbb{Q}} \mathbb{R} & \nearrow & \end{array}$$

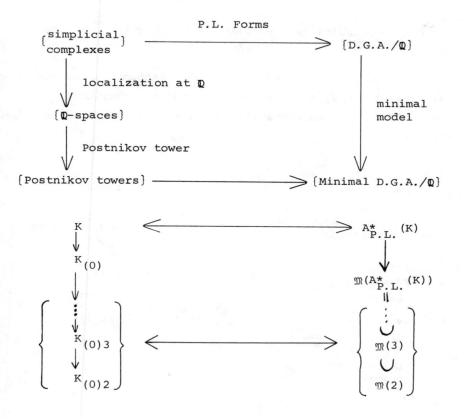

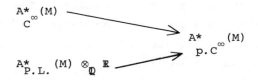

induce isomorphisms in cohomology. From this comparison theorem it follows that the minimal models satisfy

$$\mathfrak{M}(A^*_{C^\infty}(M)) \cong \mathfrak{M}(A^*_{P.L.}(M)) \otimes_{\mathbb{Q}} \mathbb{R}$$

This is the precise statement that "the deRham complex contains all the real algebraic-topological information from the manifold M". Schematically the theory is arranged as follows:

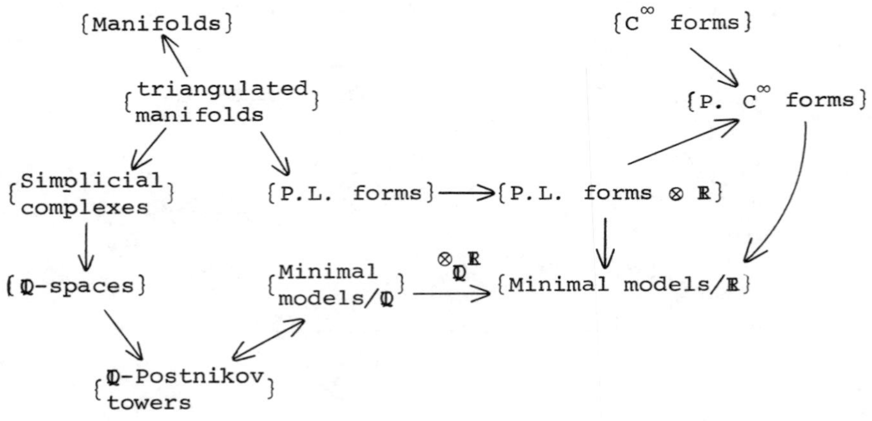

Though these notes concentrate mainly on the case of simply connected spaces, there are generalizations to the nonsimply connected case. In purely algebraic terms, part of the theory of the nonsimply connected case is similar to the simply connected one. When we try to make comparisons with homotopy the results are much weaker. The available information from the algebra of forms which is most meaningful in classical terms deals with the fundamental group. Section 12 discusses this.

I. Basic Concepts

It will suffice for the purposes of this course (and for most other situations, also) to do homotopy theory for a restricted class of spaces. These are the spaces which are homotopy-equivalent to CW complexes. All naturally encountered spaces have this property (e.g. manifolds, algebraic varieties, loop spaces on CW complexes, $K(\pi, n)$'s, etc.). Moreover for these spaces the Whitehead theorem ($f\colon X \to Y$ is a homotopy equivalence $\Leftrightarrow$ f_* is an isomorphism on homotopy groups--cf. §2 for a proof) is true. What this means is that the usual functors of homotopy theory are powerful enough to decide when two CW complexes are homotopically equivalent.

We begin with the definition of a CW complex. Let D^n, the unit n-disc, be $\{x = (x_1,\ldots,x_n) \in \mathbb{R}^n\colon \|x\|^2 \leq 1\}$ and S^{n-1} the unit (n-1) sphere be the boundary ∂D^n of D^n. (Note: in these notes we have also used the notation e^n (read n-cell) for D^n.) Given X and a continuous map $f\colon S^{n-1} \to X$, we form the adjunction space

$$X \cup_f D^n$$

which is the quotient space of the disjoint union of X and D^n when $x \in \partial D^n$ is identified with $f(x) \in X$. (Note: above we have said that f should be a continuous map; usually we shall omit the adjective continuous with the understanding that map means continuous function.) Geometrically, what we have done is attach an n-cell to X

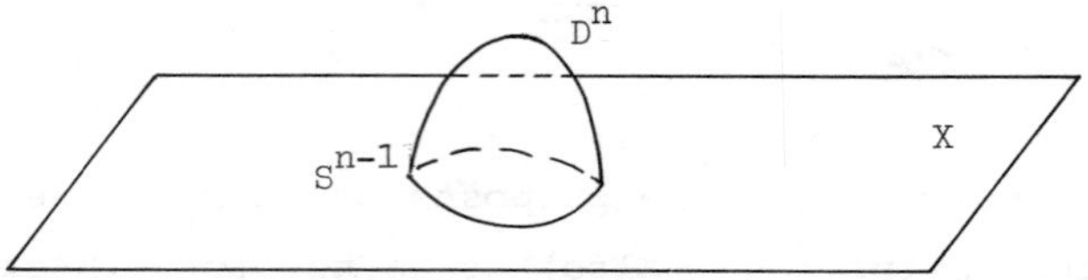

To give a space X the structure of a CW complex means intuitively that X is obtained from a point by successively attaching cells. More precisely we have subspaces $X^{(i)}$ of X with

$$\emptyset = X^{(-1)} \subset X^{(0)} \subset X^{(1)} \subset \ldots, \quad X = \cup_i X^{(i)}$$

such that (i) $X^{(i+1)}$ is obtained from $X^{(i)}$ by attaching (i+1)-cells and (ii) if $X \neq X^{(n)}$ for any n (thus X is <u>infinite dimensional</u>), then X has the <u>weak-topology</u> with respect to the $X^{(n)}$'s (i.e., $U \subset X$ is an open set $\Leftrightarrow U \cap X^{(n)}$ is open for all n). We call $X^{(n)}$ the n-<u>skeleton</u> of X. (Note: infinite dimensional CW complexes such as $\mathbb{C}P^{\infty}$, the infinite Grassmannians, the infinite sphere, etc. are very useful in homotopy theory. The weak topology means that in all cases "∞" can be well approximated by "arbitrarily large n". Thus, e.g., a map of a <u>compact</u> set into the infinite sphere S^{∞} is simply given by a map into S^N for large N, using the equatorial inclusions $S^N \subset S^{N+1} \subset \ldots$.)

<u>Examples of CW complexes</u>

(i) The n-<u>sphere</u> $S^n = \{pt.\} \cup_f D^n$ where $f: \partial D^n \to \{pt.\}$ is a degenerate attaching map.

(ii) The <u>complex projective space</u> $\mathbb{C}P^n$ is given a CW structure inductively by

$$\mathbb{C}P^n = \mathbb{C}P^{n-1} \cup_f D^{2n}$$

where

$$f: S^{2n-1} \longrightarrow \mathbb{C}P^{n-1}$$

is the Hopf map. More precisely, if we think of $\mathbb{C}P^n$ as the lines through the origin in $\mathbb{C}^{n+1}$, then taking D^{2n} to be the unit ball in $\mathbb{C}^n$, the attaching map $f: \partial D^{2n} \to \mathbb{C}P^{n-1}$ assigns to each point on the unit sphere in $\mathbb{C}^n$ the line joining that point to the origin.

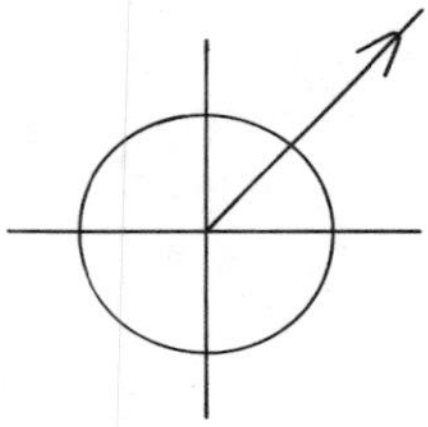

$$f(z) = \overrightarrow{0z} \in \mathbb{C}P^{n-1}$$

(iii) $\mathbb{C}P^\infty = \lim_{n\to\infty} \mathbb{C}P^n$ is the infinite CW complex having one 2n-cell for each $n \in \mathbb{Z}^+$ and with the attaching maps given as above.

(iv) Any <u>simplicial complex</u> K has the natural structure of a CW complex. The n-cells of this CW structure are exactly the n-simplices. Conversely, if X is a CW complex, then there is a simplicial complex K and a homotopy equivalence from K to X. (c.f. exercise 13.)

(v) A <u>CW-pair</u> (X,A) is a pair of spaces $A \subset X$ such that X is obtained from A by attaching cells. (It is not necessary that A itself be a CW complex.) Again, if X is obtained by attaching infinitely many cells to A, then X is given the limit (or weak) topology. If (X,A) is a CW-pair, then we denote by $X^{(n)} \cup A$ the union of A with all cells of dimension $\leq n$. If X is a CW complex and $A \subset X$ is a subcomplex, then (X,A) is a CW-pair.

CW complexes are constructed so that, almost by definition, one works inductively up through the skeleton. As an example of this we prove the Homotopy Extension Theorem for CW-pairs.

Theorem 1.1: Given a CW-pair (Y,X), a map $f: Y \to Z$, and a homotopy $F: X \times I \to Z$ from $f|X$ to $f': X \to Z$, then there is an extension $G: Y \times I \to Z$ of F such that $G(y,0) = f(y)$.

Proof: Step I. Given $f: D^n \to Z$ and $F: S^{n-1} \times I \to Z$ with $F|S^{n-1} \times \{0\} = f|\partial D^n$, find $G: D^n \times I \to Z$ extending F with $G|D^n \times \{0\} = f$.

In the picture of $D^n \times I$

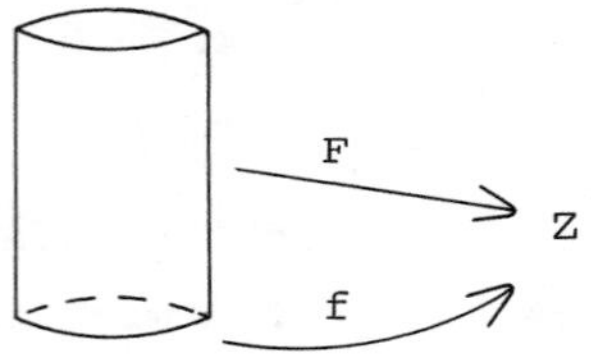

we are given a map on the "bottom" ($D^n \times \{0\}$) and the "side" ($\partial D^n \times I$). We want to extend to a map on all of $D^n \times I$. This is done by projecting $D^n \times I$ onto $(D^n \times \{0\}) \cup (S^{n-1} \times I)$ from the point $\{(\text{middle of } D^n) \times \{2\}\}$

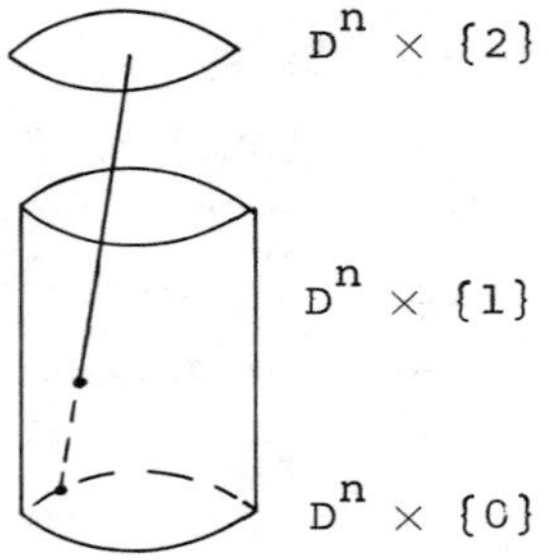

and defining $G(y,t) = G(p(y,t))$. Note: in this argument, as throughout homotopy theory, 99% of the proof is to find the correct "picture". If this is done properly, no geometric argument will be difficult (although some algebraic computations may be messy).

Step II. Given $f: Y \to Z$ and $F: X \times I \to Z$ we inductively construct $G^{(i)}: (Y \times \{0\}) \cup [(X \cup Y^{(i)}) \times I\} \to Z$. Given $G^{(i-1)}$, consider any i-cell D^i_α and attaching map $\partial D^i_\alpha \to Y^{(i-1)}$. Then we have:

$$G^{(i-1)} \circ f_\alpha \times I: (S^{i-1} \times I) \cup (D^i_\alpha \times \{0\}) \longrightarrow Z$$

and we may use Step I to extend this over $D^i_\alpha \times 1$ to a map $G^{(i)}_\alpha$. Doing this over each i-cell gives $G^{(i)}$. Let $G = \cup_i G^{(i)}$ (i.e. $G|Y^{(i)} = G^{(i)}$). By the definition of the weak topology, G is continuous and gives the required extension of F, by the construction.∎

We will always consider CW complexes modulo an equivalence relation, that of homotopy equivalence. Two maps $f_0, f_1: X \to Y$ are homotopic if there exists $F: X \times I \to Y$ with $F(x,0) = f_0(x)$ and $F(x,1) = f_1(x)$. We may set $f_t(x) = F(z,t)$ and think of the f_t as giving a continuous deformation of f_0 into f_1. Maps $f: X \to Y$ and $g: Y \to X$ are homotopy inverses if $g \circ f \sim id_X$ and $f \circ g \sim id_Y$ (here, the notation "$\sim$" means "is homotopic to", and $id_X: X \to X$ is the identity map). A map $f: X \to Y$ is a homotopy equivalence if it has a homotopy inverse; X and Y are homotopy equivalent if there is a homotopy equivalence $f: X \to Y$. Homotopy equivalence is an equivalence relation on the collection of CW complexes. In homotopy theory one considers spaces as equivalent if they are homotopy equivalent (in particular topological dimension is not defined). For the rest of this course, space will mean CW complex, with the only exception that we shall speak of path spaces and loop spaces on CW complexes (definitions below), which are not CW complexes as they stand. However, they always have the homotopy type of a CW complex (cf. Milnor, Trans. Amer. Math. Soc., vol 50 (1959) pp. 272-280) and so may be unambiguously considered as "spaces".

In many problems in homotopy theory we wish to make a construction relative to a map

$$f\colon X \longrightarrow Y.$$

It is frequently easier to work with an inclusion rather than an arbitrary map. This is always possible up to homotopy equivalence.

Theorem 1.2: *Given* $f\colon X \to Y$, *there is a space* M_f, *the mapping cylinder of* f,

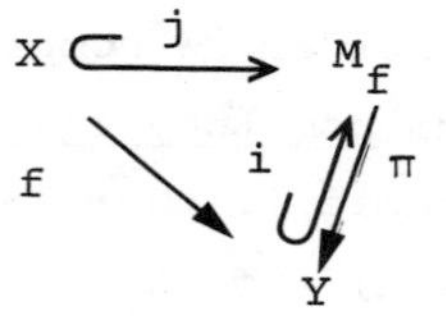

where π *and* i *are homotopy inverses and* $\pi \circ j = f$. *(Thus, we may replace* Y *by a homotopy equivalent space in which* X *is included*.)

Proof: Define $M_f = (X \times I) \cup_f Y$

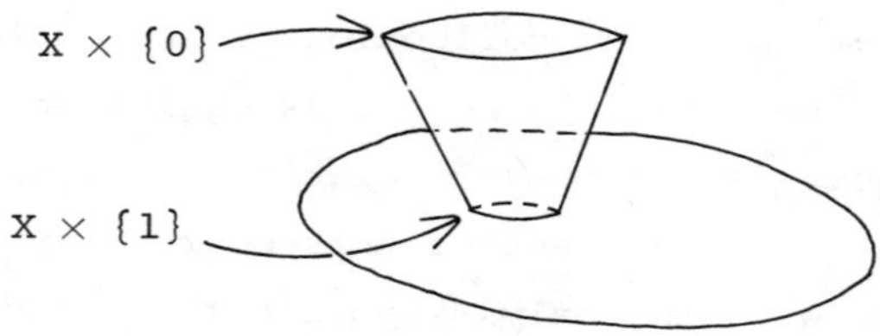

where $(x,1)$ is identified with $f(x) \in Y$. Then $\pi\colon M_f \to Y$ is given by $\pi(x,t) = f(x)$, $\pi(y) = y$ (this is consistent) and this gives a retraction of M_f onto Y. ∎

Note: If X and Y are CW complexes and $f: X \to Y$ is a cellular map (i.e. $f(X^{(i)}) \subset Y^{(i)}$), it is easy to give M_f the structure of a CW complex. In exercise (32) a proof of the fact that any f is homotopic to a cellular map f' is outlined, so that $M_f \sim M_{f'}$ which is a CW complex (the notation $A \sim B$ means that A and B are homotopy equivalent). An equivalence class of homotopy equivalent spaces is said to be a homotopy type. Thus we may consider any map as an inclusion without leaving the category of CW complexes.

Above we discussed the homotopy extension property (h.e.p.) and proved that a sub-complex of a CW complex always has the h.e.p. Now there is a dual property to the h.e.p. called the homotopy lifting property or covering homotopy property.

Given spaces E, B and a map $\pi: E \to B$ we say that $\pi: E \to B$ has the homotopy lifting property if given any space Y, a map $Y \xrightarrow{f} E$, and a homotopy g_t of $g = \pi \circ f$, there is a homotopy f_t of f such that $\pi \circ f_t = g_t$ (thus the homotopy f_t "covers" the homotopy g_t).

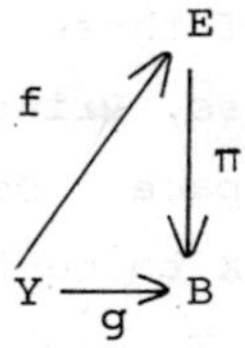

Here f is said to be a lifting of g and covering homotopy says that if a map g can be lifted, then any homotopy g_t of g can be lifted also. Not all maps have the homotopy lifting property; e.g., if B is connected, then π must be onto. If $\pi: E \to B$ has the h.l.p. (= homotopy lifting property), then it is said to be a fibration. The fiber

$F_b = \pi^{-1}(b)$ $(b \in B)$ is the inverse image of a point. Any two fibers are homotopy equivalent provided that the base is path connected (cf. Lecture 3). We let F be any space having the homotopy type of F_b (F is called a typical fiber). We write the fibration as

$$\begin{array}{ccc} F & \longrightarrow & E \\ & & \big\downarrow \pi \\ & & B \end{array}$$

having in mind a picture like

$$\begin{array}{ccc} F_b & \hookrightarrow & E \\ \big\downarrow & & \big\downarrow \pi \\ b & \hookrightarrow & B \end{array}$$

<u>Examples of fibrations</u>.

(i) Locally trivial <u>fiber bundles</u>, <u>vector bundles</u> and the associated <u>sphere bundles</u>, <u>covering spaces</u> (for a discussion of these, cf. Steenrod, The Topology of Fiber Bundles, Princeton Univ. Press).

(ii) Let X be a space. Define the <u>path space</u> $\mathcal{P}(X)$ based at $x_0 \in X$ to be the set of all paths given by maps $\omega\colon I \to X$, $\omega(0) = x_0$, and with the compact open topology. Thus a sub-basis for the open sets in $\mathcal{P}(X)$ is given by taking $K \subset I$ a compact subset, $U \subset X$ an open set, and letting $\langle K,U\rangle$ be all maps $\omega\colon I \to X$ with $\omega(K) \subset U$. Define $\pi\colon \mathcal{P}(X) \to X$ by $\pi(\omega) = \omega(1)$.

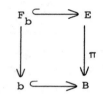

Proposition 1.3: $\pi: \mathcal{P}(X) \to X$ is a fibration.

Proof: Given a path $g: I \to X$ and an element $\tilde{g}_0 \in \mathcal{P}(X)$ such that $\pi(\tilde{g}_0) = g(0)$ (i.e., given a path g in X and a path beginning at x_0 and ending at $g(0)$), we define $\tilde{g}_t \in \mathcal{P}(X)$ by

$$\tilde{g}_t(s) = \begin{cases} \tilde{g}_0(s(1+t)) & 0 \le s \le \frac{1}{1+t} \\ g(s(1+t)-1) & \frac{1}{1+t} \le s \le 1. \end{cases}$$

One sees easily that $\pi(\tilde{g}_t) = g_t$, and that $t \to \tilde{g}_t$ is a continuous mapping of I into $\mathcal{P}(X)$. This proves the homotopy lifting property for points. One checks that the construction varies continuously with the original data, and hence gives the homotopy lifting property for all spaces.∎

Definition: The fiber $\pi^{-1}(x_0) \subset \mathcal{P}(X,x_0)$ is denoted $\Omega(X,x_0)$ and is the loop space of X based at x_0.

Lemma 1.4: $\mathcal{P}(X,x_0)$ is contractible (i.e., homotopy equivalent to a point).

Proof: Define $\mathcal{P}(X,x_0) \times I \xrightarrow{c} \mathcal{P}(X,x_0)$ by

$$c(\omega,t)(s) = \omega(ts).$$

Clearly, $c(\omega,0)$ is the constant path at x_0 for any $\omega \in \mathcal{P}(X,x_0)$, and $c(\omega,1) = \omega$.∎

We have seen how to replace any map by an inclusion up to homotopy. It is also possible to replace any map by a fibration up to homotopy. Given $f: X \to Y$, form the set of pairs $\{(x,\omega) \mid \omega(0) = f(x)\}$. This is a subspace of $X \times \mathcal{P}(Y)$

where $\mathcal{P}(Y)$ represents the space of <u>all</u> paths in Y. (Thus, $\mathcal{P}(Y) = Y^I$ with the compact-open topology.) Call this space $\tilde{X}$. Define $\pi\colon \tilde{X} \to Y$ by $\pi(x,\omega) = \omega(1)$. Define $i\colon X \hookrightarrow \tilde{X}$ by $i(x) = (x,$ constant path of $f(x))$. One checks easily that $\pi \circ i = f$, that $i\colon X \hookrightarrow \tilde{X}$ is a homotopy equivalence, and that $\pi\colon \tilde{X} \to Y$ is a fibration.∎

The freedom to convert maps into either inclusions or fibrations in homotopy category illustrates the elasticity of this category; to balance this flexibility, one has the perversity of nature that the homotopy groups of the simplest spaces, the spheres, have thus far been impossible to calculate.

Now we will introduce the language of <u>categories</u> and <u>functors</u> which so permeates mathematics today. There are two reasons for doing this. One is that it is encountered when reading reference articles, and the second is that it gives a convenient formalism for stating and remembering many of the results contained in these notes. In the notes we do not always bother to give the category-theoretic reformulation of the results we prove. These will sometimes be left to the reader as exercises.

A <u>category</u>, $\mathcal{C}$, consists of a collection of objects $\mathrm{Obj}(\mathcal{C})$ and <u>morphisms</u> between two objects of $\mathcal{C}$, A and B. We will require that the collection of all morphisms from A to B be a set (though this is unnecessary). This set is denoted by $\mathrm{Hom}(A,B)$ or $\mathrm{Hom}_{\mathcal{C}}(A,B)$. Several axioms are required to be satisfied.

1. $\mathrm{Hom}(A,A)$ always contains a distinguished element I_A, the identity morphism of A.

2. There is a composition $\mathrm{Hom}(A,B) \times \mathrm{Hom}(B,C) \to \mathrm{Hom}(A,C)$ (A, B, C objects of $\mathcal{C}$) which is associative $(fg)h = f(gh)$ and for which I_A is a unit; i.e., $f \circ I_A = f$ and $I_B \circ f = f$ for any $f \in \mathrm{Hom}(A,B)$.

Note: The objects of $\mathcal{C}$ do <u>not</u> necessarily have to be sets and a morphism from A to B does not have to be a set map. It is just considered as an arrow.

<u>Examples</u>: 1. <u>Abelian groups</u>. The objects are abelian groups, the morphisms are group homomorphisms.

2. <u>Topological spaces</u>. the objects are the usual topological spaces, the morphisms are continuous maps.

3. <u>Homotopy category</u>. The objects are spaces homotopy equivalent to CW complexes and the morphisms are homotopy classes of maps.

4. Analogous to 1) we have the categories of <u>groups</u>, <u>rings</u>, <u>fields</u>, <u>vector spaces over a field</u>, and <u>modules over a ring</u>.

An <u>isomorphism</u> in a category is a morphism with an inverse; i.e., $f: A \to B$ is an isomorphism if there exists $g: B \to A$ with $g \circ f = I_A$ and $f \circ g = I_B$.

The isomorphisms in 1,2, and 3 are respectively group isomorphism, homeomorphism, and homotopy equivalence.

A <u>functor</u> between two categories, $\mathfrak{F}: \mathcal{C} \to \mathcal{C}'$, is an assignment of objects $\mathfrak{F}(\mathfrak{A})$ in $\mathcal{C}'$ to objects $\mathfrak{A}$ in $\mathcal{C}$ and morphisms $\mathfrak{F}(f)$ in $\mathrm{Hom}_{\mathcal{C}'}(\mathfrak{F}(\mathfrak{A}), \mathfrak{F}(\mathfrak{B}))$ to f in $\mathrm{Hom}_{\mathcal{C}}(\mathfrak{A}, \mathfrak{B})$ such that identities and compositions are preserved. Clearly a functor sends isomorphic objects in $\mathcal{C}$ to isomorphic objects in $\mathcal{C}'$.

A <u>homotopy functor</u> is a functor from the homotopy category to another category. <u>Algebraic topology</u> is the study of homotopy functions into algebraic categories--i.e. groups, rings, fields, chain complexes, vector spaces, etc. Under a homotopy functor, homotopy equivalent spaces have assigned to them isomorphic objects. This is why, in homotopy theory, it is permissible to identify homotopy equivalent objects.

As an example of a functor of algebraic topology consider

homology. The homology functor assigns to each space X a graded group $H_*(X)$, and to a continuous map $f\colon X \to Y$ an induced map on homology $f_*\colon H_*(X) \to H_*(Y)$. If $f \sim g$, then $f_* = g_*$ so that homology is actually a homotopy functor.

The singular homology groups of X are obtained from the chain complex

$$\xrightarrow{\partial_{n+1}} C_n(X) \xrightarrow{\partial_n} C_{n-1}(X) \xrightarrow{\partial_{n-1}} \cdots \xrightarrow{\partial_1} C_0(X)$$

where $C_n(X)$ is the free abelian group generated by all continuous maps $f\colon\Delta^n \to X$ of the standard n-simplex Δ^n into X and

$$\partial_n(f) = \Sigma(-1)^i f|(\text{i-face of } \Delta^n).$$

One checks combinatorially that $\partial\partial = 0$, and then defines $H_n(X) = \operatorname{Ker} \partial_n/\operatorname{Im} \partial_{n+1}$.

We may define homology with coefficients in an abelian group G by forming

$$\cdots \longrightarrow (C_n(X) \otimes G \xrightarrow{\partial_n \otimes 1} C_{n-1}(X) \otimes G \longrightarrow \cdots \longrightarrow C_0(X) \otimes G$$

and taking its homology. These groups are denoted $H_n(X;G)$. When no coefficients are specified, $\mathbb{Z}$ coefficients are understood.

We also have cohomology functors. The cohomology of X is obtained from the singular cochain complex

$$\xleftarrow{\delta_{n+1}} \operatorname{Hom}(C_n(X),G) \xleftarrow{\delta_n} \operatorname{Hom}(C_{n-1}(X),G) \leftarrow \cdots \leftarrow \operatorname{Hom}(C_0(X),G))$$

$$(\delta_n = \operatorname{Hom}(\partial_n, 1))$$

by taking $\operatorname{Ker} \delta_{n+1}/\operatorname{Im} \delta_n$.

These groups are denoted $H^n(X,G)$. Again if no G is

made explicit, $\mathbb{Z}$ coefficients are understood.

$H^n(X)$ has the structure of a <u>graded-commutative ring</u>. That is to say, there is a multiplication $H^n(X) \otimes H^m(X) \overset{\cup}{\to} H^{n+m}(X)$ with the property that $\alpha \cup \beta = (-1)^{\dim \alpha \cdot \dim \beta} \beta \cup \alpha$. This cup product is induced from one on the cochain level given by

$$\langle \alpha_p \cup \beta_q, \Delta^{p+q} \rangle$$

$$= \langle \alpha_p, \text{front } p\text{-face of } \Delta^{p+q} \rangle \cdot \langle \beta_q, \text{back } q\text{-face} \rangle.$$

It is <u>not</u> true on the cochain level that $\alpha_p \cup \beta_q = (-1)^{pq} \beta_q \cup \alpha_p$. But one is able to show that the cup product on the cochain level induces one on the cohomology level which is, in fact, graded-commutative.

For a good introduction to CW complexes, homology, and cohomology consult Greenberg's "Lectures on Algebraic Topology". For a more encyclopedic treatise on algebraic topology which covers all the homotopy theory presented in this course, save localization, one should see Spanier's "Algebraic Topology". For another account of some of the topics presented later in this course, such as obstruction theory, one should see Hu's book "Elementary Homotopy Theory".

II. CW Homology Theorem

If X is a space, $H_*(X)$ denotes the singular homology of X with $\mathbb{Z}$-coefficients. We recall briefly the construction and basic properties of $H_*(X)$.

Let Δ^n be the standard n-simplex in $\mathbb{R}^{n+1}$, and let $\mathrm{Sing}_n(X)$ be the free abelian group generated by the singular n-simplices

$$\Delta^n \xrightarrow{f} X ,$$

which by definition are the continuous maps of Δ^n to X. The boundary operator

$$(x_0,\ldots,x_n) \longmapsto \sum_{i=0}^{n} (-1)^i (x_0,\ldots,\hat{x}_i,\ldots,x_n)$$

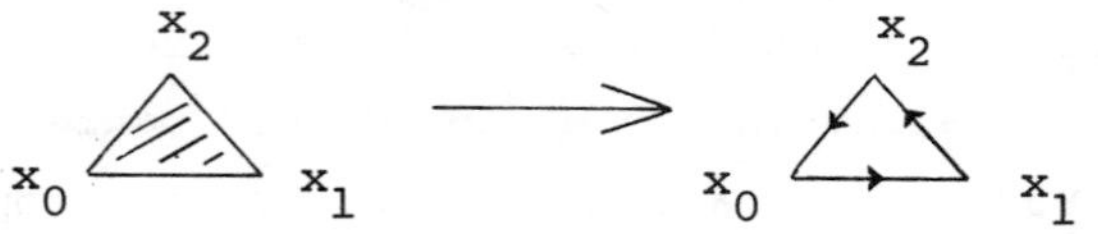

induces

$$\mathrm{Sing}_n(X) \xrightarrow{\partial} \mathrm{Sing}_{n-1}(X)$$

$$\partial \circ \partial = 0 \ .$$

The cycles $Z_n(X)$ are the singular chains $c \in \mathrm{Sing}_n(X)$ with $\partial c = 0$, and the homology is defined by

$$H_n(X) = Z_n(X)/\partial \ \mathrm{Sing}_{n+1}(X)$$

The basic properties of homology are:

(i) A map $f\colon X \to Y$ induces a map on homology

$$f_*: H_*(X) \longrightarrow H_*(Y)$$

which depends only on the homotopy class of f.

(ii) An orientation for S^n determines an isomorphism $H_n(S^n) \overset{\sim}{\to} \mathbb{Z}$. If we reverse the orientation, then this isomorphism is multiplied by -1. A representative cycle for the class corresponding to $1 \in \mathbb{Z}$ is determined by a homeomorphism $\partial\Delta^{n+1} \to S^n$ which is orientation preserving.

(iii) Given $Y \subset X$, there are defined <u>relative homology groups</u> $H_*(X,Y)$, and there is an <u>exact homology sequence</u>

$$\cdots \quad H_n(Y) \longrightarrow H_n(X) \longrightarrow H_n(X,Y) \xrightarrow{\partial} H_{n-1}(Y) \longrightarrow \cdots$$

(iv) If $U \subset Y \subset X$ is such that

$$\bar{U} \subset \text{interior}(Y),$$

then we have the <u>excision property</u>

$$H_*(X-U, Y-U) \cong H_*(X,Y).$$

(v) Any <u>homology theory</u>, that is a functor from the homotopy category to abelian groups which satisfies (ii)-(iv), is necessarily singular homology (cf. exercise (40)). The most interesting of the axioms is probably excision, which is essentially the same as Mayer-Vietoris.

Suppose now that X is a CW complex and let $X^{(n)}$ denote its n-skeleton. Then the n+1 skeleton $X^{(n+1)}$ is obtained by attaching n+1 cells to $X^{(n)}$ by maps

$$\partial e^{n+1} \xrightarrow{f} X^{(n)}.$$

Denote by $C'_n(X)$ the free abelian group generated by the oriented n-cells of X. Since each n-cell has two orientations, there are 2 generators in $C'_n(X)$ for each n-cell. Introduce the relation that the 2 generators corresponding to a cell are negatives of one another. Denote the quotient group by $C_n(X)$. It is a free abelian group of rank equal to the number of n-cells.

If we are given an orientation for an n-cell e^n of X, then choosing an orientation preserving isomorphism $\Delta^n \to e^n$ determines an element in $H_n(X^{(n)},X^{(n-1)})$. The resulting element depends only on the orientation of e^n and changes sign if we reverse the orientation. Thus the construction determines a map $\varphi_n: C_n(X) \to H_n\ X^{(n)},X^{(n-1)}$. As we shall see below an easy application of the excision theorem for homology shows that this mapping is an isomorphism. Let us assume for the moment that φ_n is an isomorphism. Define the boundary map $\partial: C_n(X) \to C_{n-1}(X)$ to be the composition:

$$C_n(X) \xrightarrow{\varphi_n} H_n(X^{(n)},X^{(n-1)}) \xrightarrow{\partial} H_{n-1}(X^{(n-1)}) \xrightarrow{i_*} H_{n-1}(X^{(n-1)},X^{(n-2)})$$

$$\xrightarrow{\varphi_{n-1}^{-1}} C_{n-1}(X).$$

<u>Theorem 2.1 (CW homology theorem)</u>: $\{C_*(X),\partial\}$ <u>is a chain complex</u> <u>There is a natural identification of its homology with the singular homology of</u> X.

<u>Remarks:</u> 1) In case X has only finitely many cells in each dimension the chain groups $C_n(X)$ are finitely-generated, free abelian groups. Hence, the above theorem gives a fairly

effective way of calculating homology.

2) If (X,A) is a CW pair, i.e., if X is built by inductively attaching cells to A, then one has an analogously defined chain complex $\{C_n(X,A),\partial\}$ which computes the relative singular homology $H_*(X,A)$.

Let us begin the proof of 2.1 with a couple of simple lemmas.

Lemma 2.2: Let $X = Y \cup_f e^m$. Then

$$\text{and} \quad \begin{cases} H_i(X,Y) = 0 & i \neq m, \\ H_m(X,Y) \cong \mathbb{Z}. \end{cases}$$

Proof: Let N be a neighborhood of ∂e^m and $U = Y \cup_f N$.

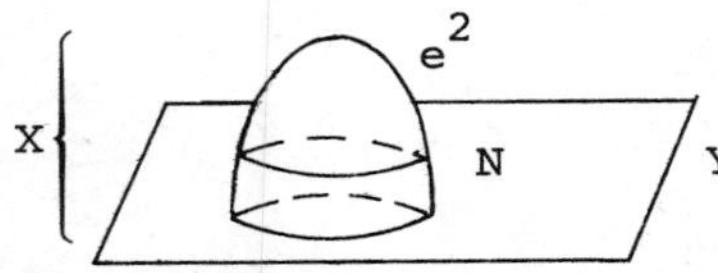

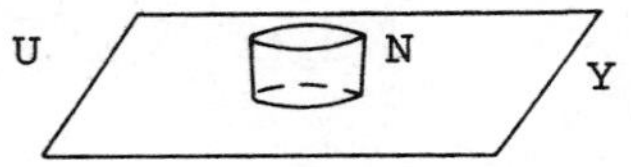

Then there is a deformation retraction of U onto Y (cf. the picture). The excision and homotopy axioms give

$$H_i(X,Y) \underset{\text{homotopy}}{\cong} H_i(X,\bar{U}) \underset{\text{excision}}{\cong} H_i(e^m - N, \partial N)$$

$$\underset{\text{homotopy}}{\cong} H_i(e^m, \partial e^m).$$

But $H_i(e^m) = 0$ for $i > 0$, and so the exact homology sequence and computation of the homology of $S^{m-1} = \partial e^m$ give

$$H_i(e^m, \partial e^m) \cong H_{i-1}(\partial e^m) \qquad (i \neq 0)$$

$$= \begin{cases} \mathbb{Z} \text{ for } i = m \\ 0 \quad \text{otherwise} \end{cases}$$ ∎

Remark: A similar proof gives the following useful result:

Proposition 2.3: Let (X,A) be a CW-pair. Let X/A denote the CW complex obtained by shrinking A to a point. Then $H_i(X,A) \cong \tilde{H}_i(X/A)$ where $\tilde{H}_*$ denotes the reduced homology.

The proof is by induction on the dimension of the cells and proceeds similarly to the proof of 2.2. Details are left to the reader.

It is excision which makes homology more computable than other homotopy functors such as the homotopy groups.

Lemma 2.4: $H_i(X,X^{(n-1)}) = 0 \qquad \text{for } i \leq n-1$

$$H_i(X^{(n)},X^{(n-1)}) = 0 \qquad \text{for } i \neq n.$$

The map $\varphi_n: C_n(X) \to H_n(X^{(n)},X^{(n-1)})$ is an isomorphism.

Proof: These statements are all proved from Lemma 2.2 using induction on the number of cells (plus a direct limit argument if X has infinitely many cells). ∎

Proof of 2.1: We have an isomorphism:

$$\varphi_n: C_n(X) \longrightarrow H_n(X^{(n)},X^{(n-1)})$$

and $\partial: C_n(X) \to C_{n-1}(X)$ is $\varphi_{n-1}^{-1} \circ \tilde{\partial} \circ \varphi_n$ where $\tilde{\partial}: H_n(X^{(n)},X^{(n-1)}) \to H_{n-1}(X^{(n-1)},X^{(n-2)})$ is the boundary in the homology long exact sequence for the triple

$X^{(n-2)} \subset X^{(n-1)} \subset X^{(n)}$. Since $\tilde{\partial}_{n-1} \circ \tilde{\partial}_n = 0$ it follows that $\partial \circ \partial \colon C_n(X) \to C_{n-2}(X)$ is zero. Thus, $\{C_*(X), \partial\}$ is a chain complex. Applying Lemma 2.2 we have an exact sequence

$$\begin{array}{ccccc}
0 \to H_n(X^{(n)}, X^{(n-2)}) & \to & H_n(X^{(n)}, X^{(n-1)}) & \xrightarrow{\tilde{\partial}} & H_{n-1}(X^{(n-1)}, X^{(n-2)}) \\
\wr\| & & \wr\| & & \wr\| \\
0 \to H_n(X^{(n)}) & \longrightarrow & C_n(X) & \xrightarrow{\partial} & C_{n-1}(X)
\end{array} .$$

Thus, $H_n(X^{(n)})$ is identified with the cycles in $C_n(X)$. Similarly, there is a commutative diagram:

$$\begin{array}{ccc}
C_{n+1}(X) = H_{n+1}(X^{(n+1)}, X^{(n)}) & \xrightarrow{\partial'} & H_n(X^{(n)}) \\
\downarrow & \nearrow \partial & \\
H_{n+1}(X, X^{(n)}) & & \\
\downarrow & & \\
0 & &
\end{array}$$

This allows us to identify $\partial C_{n+1}(X) \subset Z_n(C_*(X))$ with $\partial(H_{n+1}(X, X^{(n)})) \subset H_n(X^{(n)})$. Applying 2.2 and 2.4 once again we find the following is exact:

$$H_{n+1}(X, X^{(n)}) \xrightarrow{\partial} H_n(X^{(n)}) \longrightarrow H_n(X) \longrightarrow 0.$$

Thus, $Z_n(C_*(X))/\partial C_{n+1}(X) = H_n(X^{(n)})/\partial(H_{n+1}(X, X^{(n)})) = H_n(X)$.

If $f\colon X \to Y$ is a continuous mapping between CW complexes, then f is homotopic to a cellular map. If we deform f to be cellular then it induces maps on all the relative groups used in establishing the equivalent of $H_*(C_*(X))$ with $H_*(X)$. Hence this equivalence is functorial.∎

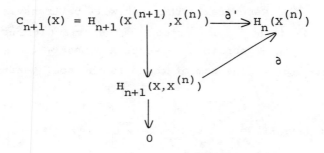

<u>Remark</u>: Choose an orientation for each n-cell and each (n+1)-cell of X. If e_α^{n+1} is an (n+1)-cell, then $\partial e_\alpha^{n+1} = \Sigma_\beta\, c_{\alpha\beta} e_\beta^n$ for some coefficients $c_{\alpha\beta}$. This means that if $f_\alpha\colon \partial e_\alpha^{n+1} \to X^{(n)}$ is the attaching map for e_α^{n+1}, then there is a singular chain c in $X^{(n-1)}$ so that $\partial e_\alpha^{n+1} - \Sigma\, c_{\alpha\beta} e_\beta^n - c$ is a singular boundary in $X^{(n)}$.

If the attaching maps are generic, then there is an appealing geometric description of $\partial\colon C_n(X) \to C_{n-1}(X)$. Choose an oriented n-cell $(e^n, \mathcal{O}_e)$ and an oriented (n-1)-cell $(\sigma^{n-1}, \mathcal{O}_\sigma)$. To calculate the coefficient of $(\sigma^{n-1}, \mathcal{O}_\sigma)$ in $\partial(e^n, \mathcal{O}_e)$ one simply considers the attaching map $\partial e^n \xrightarrow{\varphi_e} X^{(n-1)}$. Generically this map will be transverse to some point p_σ in σ^{n-1}. Comparison of the orientations allows us to assign an algebraic intersection number ± 1 to each point in $\varphi_e^{-1}(p_\sigma)$. The sum of these is the sought-after coefficient.

<u>Examples</u>: (i) $\mathbb{R}P^2 = S^1 \cup_{z^2} D^2$ where $D^2 = \{z \in \mathbb{C}\colon |z| \le 1\}$

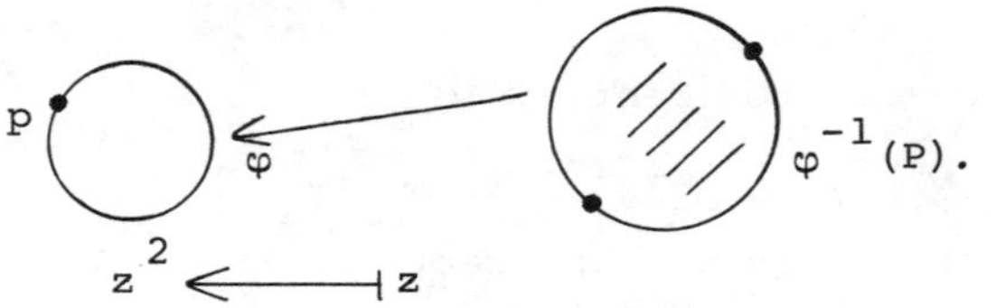

The boundary map sends $[D^2, \mathcal{O}_{D^2}] \to 2[e^1, \mathcal{O}_{e^1}]$.

(ii) The 2-torus is obtained by attaching D^2 to $S^1 \vee S^1$ by $aba^{-1}b^{-1}$

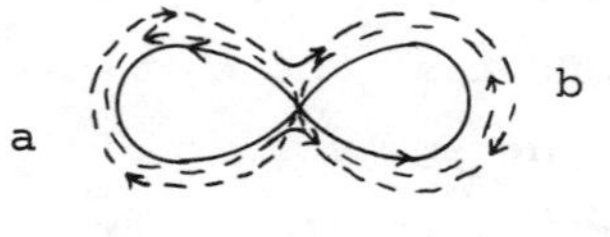

$\varphi^{-1}(P_a)$ = 2 points with opposite sign

$\varphi^{-1}(P_b)$ = 2 points with opposite sign.

Hence $\partial[D^2, \mathcal{O}_{D^2}] = 0$.

(iii) The n-sphere $S^n = x_0 \cup_f e^n$. Thus, $H_i(S^n) = 0$ for $i \neq 0,n$ and $H_n(S^n) = \mathbb{Z}$.

(iv) Complex projective space $\mathbb{C}P^n$ is obtained from $\mathbb{C}P^{n-1}$ by attaching a cell of dimension 2n. Inductively, one argues that $\mathbb{C}P^n$ has a cell decomposition with exactly one cell occuring in dimensions 0, 2, 4,...,2n. Thus, in $C_*(X)$ all boundary maps must be zero. Hence,

$$H_{2i}(\mathbb{C}P^n) = \mathbb{Z} \qquad \text{for} \quad 0 \leq i \leq n$$

$$H_*(\mathbb{C}P^n) = 0 \qquad \text{for all other } *.$$

(v) Consider the map $f_n: S^1 \to S^1$ given by $e^{i\theta} \to e^{in\theta}$ for $n \in \mathbb{Z}$. Let $X_n = S^1 \cup_{f_n} e^2$. Then

$$H_1(X_n) = \mathbb{Z}/n\mathbb{Z}.$$

The reason is that $\partial e^2 = nS^1$ (cf. example 1 for n = 2).

(vi) As we saw in §I a simplicial complex has a natural CW complex structure. The CW homology theorem gives a purely combinatorial way to calculate $H_*(|K|)$. (Here $|K|$ is the topological space associated to K.) The complex $\{C_*(|K|),\partial\}$ is exactly the simplicial chain complex.

(vii) If (X,A) is a CW-pair, then there is an analogously defined complex $\{C_n(X,A),\partial\}$. The group $C_n(X,A)$ is a free abelian group with one generator for each n-cell in X-A. Arguments similar to the ones above show that the homology of this complex is naturally identified with $H_*(X,A)$.

III. The Whitehead Theorem and the Hurewicz Theorem

A. Definitions and elementary properties of homotopy groups.

If X and Y are spaces, then $[X,Y]$ denotes the set of homotopy classes of maps from X to Y. If (X,A) and (Y,B) are pairs, then $[(X,A),(Y,B)]$ denotes the set of homotopy classes of maps of the pair (X,A) to the pair (Y,B). If H is such a homotopy, then for each t H_t is required to map A to B. If (X,x_0) and (Y,y_0) are based spaces then we denote $[(X,x_0),(Y,y_0)]$ by $[X,Y]_0$.

Recall that the set $[S^1,X]_0$ has a group structure induced by composing loops. This group is the fundamental group and is denoted $\pi_1(X,x_0)$ (or $\pi_1(X)$ if the base point plays no important role). We define sets $\pi_n(X,x_0) = [S^n,X]_0$. Then $\pi_0(X)$ is exactly the set of path components of X. We show how to make $\pi_n(X)$ an abelian group for $n \geq 2$.

To define the group structure on $\pi_n(X)$, $n \geq 2$, and show that it is abelian it is convenient to give a different description. Let I^n denote the n cube. Then $\pi_n(X) = [(I^n,\partial I^n);(X,x_0)]$. The composition is

$f \cdot g =$ 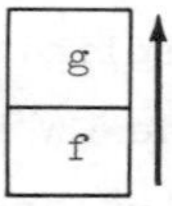

1st direction.

This is abelian as the following sequence of pictures indicates (the unmarked areas are sent to the base point).

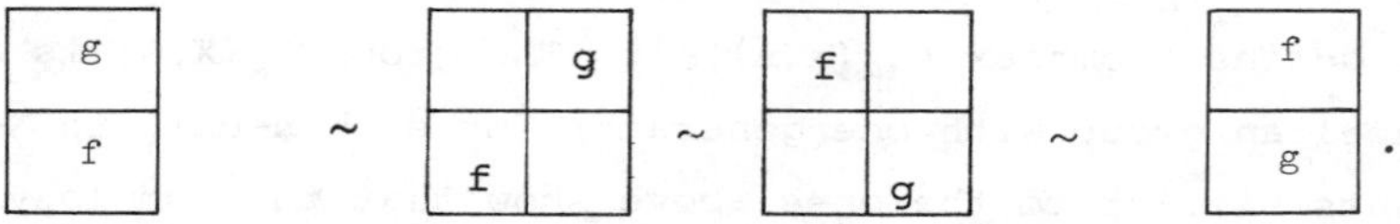

For $n \geq 2$ define $\pi_n(X,A)$ to be

$$[(I^n, \partial I^n, \overline{\partial I^n - I^{n-1} \times \{1\}}), (X,A,a_0)].$$

The group law is given as before. An argument similar to the one above shows that for $n \geq 3$, $\pi_n(X,A)$ is an abelian group. Given $f \in [(X,A),(Y,B)]$, there is an induced map

$$f_*: \pi_*(X,A) \longrightarrow \pi_*(Y,B).$$

This means that the homotopy groups form a covariant functor from the homotopy category (of pairs) of spaces to the category of groups.

Remarks: (i) The homotopy groups are easier to define but much harder to calculate than the homology groups. One of the main points of this course will be to give an effective way to calculate $\pi_* \otimes \mathbb{Q}$.

(ii) We show that $\pi_k(S^n) = 0$ if $k < n$ and $\pi_3(S^2) \neq 0$. To see the first let $f: S^k \to S^n$ be given with $k < n$. Deform f until it is cellular. Since the k-skeleton of S^n can be taken to be a point, this deformation carries f to a constant map. (In general, this argument shows that $\pi_k(X^{(m)}) \xrightarrow{\sim} \pi_k(X)$ for $k \leq m-1$, and that $\pi_k(X^{(k)}) \to \pi_k(X)$ is onto.) To show that $\pi_3(S^2) \neq 0$ consider $\mathbb{C}P^2$ as an adjunction space $\mathbb{C}P^2 = (\mathbb{C}P^1) \cup_f e^4$ where $\mathbb{C}P^1 = S^2$. The homotopy type of an adjunction space depends only on the homotopy class of the attaching map. Thus, if $\pi_3(S^2) = 0$, then $\mathbb{C}P^2$ would be homotopy equivalent to $S^2 \cup_{\text{constant}} e^4 = S^2 \vee S^4$. This would imply that if g generates $H^2(\mathbb{C}P^2)$, then $g \cup g = 0$ in $H^4(\mathbb{C}P^2)$. But we know (say by Poincaré duality) that $g \cup g \neq 0$.

There is an anlogous argument showing that $\pi_{4n-1}(S^{2n}) \neq 0$.

If $f\colon S^n \to X$, then there is induced $f_*\colon H_n(S^n) \to H_n(X)$. But $H_n(S^n)$ is identified, once and for all, with $\mathbb{Z}$. Thus we have $f_*(1) \in H_n(X)$. This determines a natural transformation:

$$H\colon \pi_n(X) \longrightarrow H_n(X)$$

$$[f] \longrightarrow f_*(1)$$

called the <u>Hurewicz homomorphism</u>. (It is an easy exercise to show that H is indeed a homomorphism.) Applying this in the special case where $X = S^n$ we have

$$H\colon \pi_n(S^n) \longrightarrow H_n(S^n) = \mathbb{Z}.$$

$H(f)$ is also called the <u>degree of</u> f, and some times denoted by $\deg(f)$.

<u>Theorem 3.1 (Brouwer)</u>: <u>The map</u> $H\colon \pi_n(S^n) \to \mathbb{Z}$ <u>is an isomorphism</u>.

Since $H(\mathrm{Id}_{S^n}) = 1$, it is obvious that H is onto. It is harder to show that H is one-to-one, i.e. that if $\deg(f) = \deg(g)$, then $f \simeq g$. We shall prove this later in this section.

If $x_0 \in A \subset X$, then there is an exact homotopy sequence

$$\cdots \longrightarrow \pi_{n+1}(X,A) \xrightarrow{\partial} \pi_n(A) \xrightarrow{i} \pi_n(X) \xrightarrow{j} \pi_n(X,A) \longrightarrow \cdots$$

As one consequence of this and the calculations above we see that:

<u>Excision is false for the homotopy groups</u>.

The easiest example of this is that $\pi_3(D^2,S^1) \overset{\partial}{\underset{\approx}{\to}} \pi_2(S^1)$ is zero whereas $\pi_3(S^2,D^2) \cong \pi_3(S^2) \neq 0$. Of course, were excision to hold for homotopy groups, then they would satisfy all the axioms for homology. If that were true, then the argument in Section 2 which identified singular homology with $H_*(C_*(X,A))$ could be used to identify the homotopy groups with $H_*(C_*(X,A))$. It would follow that the homology and homotopy groups were isomorphic.

B. <u>The Whitehead Theorem</u>.

<u>Statement</u>: <u>Let</u> X <u>and</u> Y <u>be CW complexes with base points being</u> 0-<u>cells. Let</u> $f: (X,x_0) \to (Y,y_0)$ <u>be a map inducing isomorphisms</u>

$$f_*: \pi_n(X,x_0) \longrightarrow \pi_n(Y,y_0).$$

<u>Suppose that</u> Y <u>is connected. Then</u> $f: X \to Y$ <u>is a homotopy equivalence</u>.

<u>Proof</u>: Let us begin with a special case. Suppose $\dim X < \infty$ and $\pi_n(X) = 0$ for all $n \geq 0$. We shall show that X is contractible. (This is a special case of the theorem for $x_0 \hookrightarrow X$.) Since $\pi_0(X)$ is the one point set, the zero skeleton can be deformed to $x_0 \in X$.

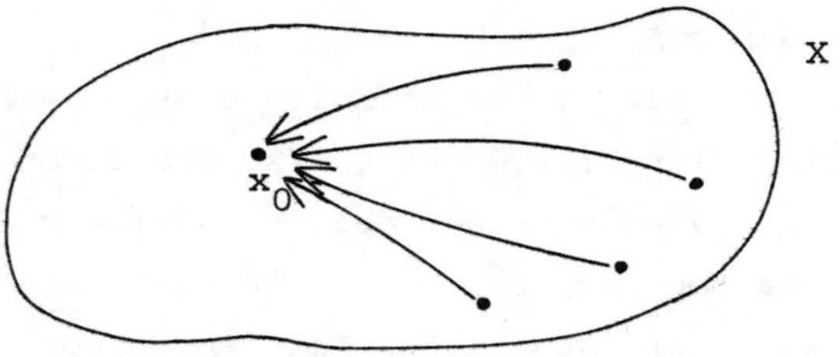

Use the homotopy extension property (Theorem 1.1) to obtain a continuous family:

$$f_t: X \longrightarrow X \quad 0 \leq t \leq 1$$

such that $f_0 = Id_X$ and $f_1(X^{(0)}) = x_0$.

Now consider $X^{(1)}$. The image under f_1 of any 1-cell, e^1, is a loop based at x_0. Since $\pi_1(X) = 0$ we can deform $f_1(e^1)$ through loops based at x_0 to the constant loop. Doing this for each 1-cell defines a homotopy from $f_1|X^{(1)}$ to the constant map $f_2: X^{(1)} \to x_0$. It is a homotopy relative to $X^{(0)}$. Using homotopy extension we can find a homotopy $f_t: X \to X$, $1 \leq t \leq 2$, with $f_2|X^{(1)}$ = constant. Continue in this fashion using the fact that $\pi_n(X) = 0$ for all n. In the end we have a homotopy from Id_X to the constant map of X to $x_0 \in X$.

<u>Remarks</u>: (i) The same argument shows that if $\pi_n(X) = 0$ for $n < N$, then there is a homotopy

$$f_t: X \longrightarrow X; \quad 0 \leq t \leq 1$$

with

$$f_0 = Id$$

and

$$f_1(X^{(N-1)}) = x_0.$$

As a consequence $\tilde{H}_k(X) = 0$ for $k < N$.

(ii) In case X is infinite dimensional, we make the first homotopy last for $0 \leq t \leq 1/2$, the second last for $1/2 \leq t \leq 3/4$, etc. and then define $f_1(X) = x_0$. Using the fact that the topology on X is the weak (or limit) topology, it is an easy exercise to show that the proposed homotopy is indeed continuous.

(iii) There is a relative version of the theorem which states that, if (X,A) is a CW-pair and if $\pi_n(X,A) = 0$ for all n, then there is a deformation retraction

$$f_1: X \longrightarrow A,$$

i.e., there is a homotopy $f_t: X \to X$; $0 \leq t \leq 1$, such that f_0 = identity, $f_1(X) \subset A$, and $f_t|A$ = identity. The proof is essentially the same as in the absolute case.

<u>General case</u>: Given $f: X \to Y$, consider the mapping cylinder M_f. Since M_f retracts to Y it follows that $\pi_k(M_f) = \pi_k(Y)$ for all k. If $f_*: \pi_k(X) \to \pi_k(Y)$ is an isomorphism for all k, then so is $\pi_k(X) \xrightarrow{i_*} \pi_k(M_f)$. Hence $\pi_k(M_f, X) = 0$ for all k. Thus remark (iii) implies that there is a retraction $r: M_f \to X$. The composition $Y \subset M_f \xrightarrow{r} X$ is the required homotopy inverse to f. This completes the proof of the Whitehead Theorem.

<u>Example</u>: Think of S^k as $I^k/\{\partial I^k\}$. There is a product map

$$I^2/\{\partial I^2\} \times I^2/\{\partial I^2\} \longrightarrow I^4/\{\partial I^4\}.$$

This is a map $f: S^2 \times S^2 \to S^4$. Since $\pi_k(S^2 \times S^2) \cong \pi_k(S^2) \times \pi_k(S^2)$, one sees easily that $f_*: \pi_k(S^2 \times S^2) \to \pi_k(S^4)$ is trivial for all k. But f is <u>not</u> homotopic to a constant map. The easiest way to see this is to note that $f_*: H_4(S^2 \times S^2) \to H_4(S^4)$ is nonzero (in fact it is an isomorphism).

In dealing with spaces which are not CW complexes a map $f: X \to Y$ is not necessarily a homotopy equivalence if it induces an isomorphism in all homotopy groups. In this case we say that f is a weak homotopy equivalence. This notion generates an equivalence relation weak homotopy equivalence which when restricted to CW complexes is homotopy equivalence by Whitehead's Theorem. In general if A is a CW complex and $f: X \to Y$ is a weak homotopy equivalence then $f_{\#}: [A,X] \to [A,Y]$ is a bijection.

C. Proof of the Brouwer Theorem.

Since we have seen that $H: \pi_n(S^n) \tilde{\to} \mathbb{Z}$ is onto, it suffices to show that if $H(f) = 0$, then $f \sim$ constant. First deform f until it is a C^∞ mapping and $p \in S^n$ is a regular value for f.

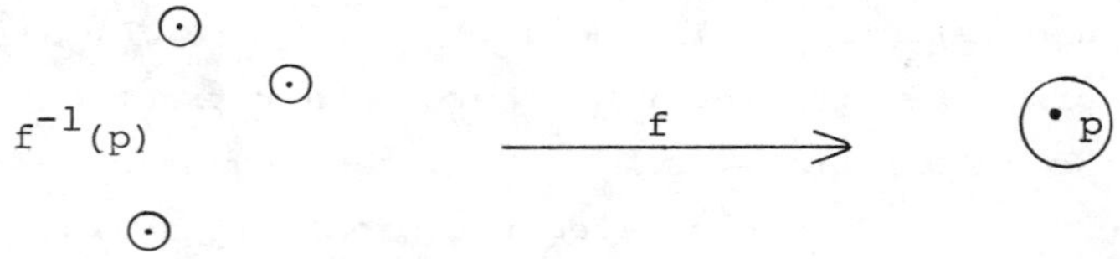

At each point $x_i \in f^{-1}(p)$ there is a local degree of f. This is +1 if f is orientation-preserving near x_i and -1 if it is orientation reversing there.

We claim that $\deg(f) = \Sigma_{x_i \in f^{-1}(p)}$ (local degree of f at x_i). The easiest way to see this is to choose a C^∞-form ω on S^n of degree n supported in a small ball $D^n \subset S^n$ containing p whose integral over D^n is 1. Clearly, $\int_{S^n} f^*\omega = \deg(f)$. On the other hand, if we choose D^n sufficiently small there will be one component of $f^{-1}(D^n)$ for each $x_i \in f^{-1}(p)$. If C_i is the component of $f^{-1}(D)$ containing x_i, then $\int_{C_i} f^*(\omega) = (\deg(f) \text{ at } x_i)$. $\int_{D^n}\omega = \deg(f)$ at x_i. The result follows immediately.

This argument assumes deRham's theorem from §VIII. Alternatively one may use

$$H_n(S^n) \cong H_n(S^n,p) \cong H_n(S^n,D^n)$$

$$\cong H_n(S^n - D^n, \partial D^n) \cong H_{n-1}(S^{n-1})$$

together with

$$H_n(S^n) \cong H_n(S^n, f^{-1}(p)) \cong H_n(S^n, f^{-1}(D^n))$$

$$\cong H_n(S^n - f^{-1}(D^n), f^{-1}(\partial D^n)) \hookrightarrow \oplus_i H_{n-1}(\partial C_i)$$

to show that

$$\deg f = \Sigma_i \deg(f|\partial C_i).$$

Thus, we must show that if $\Sigma_{x_i \in f^{-1}(p)} (\deg(f) \text{at } x_i) = 0$, then f is homotopic to a constant. We do this by induction on n, beginning with the case $n = 2$ (the case $n = 1$ will be proved afterward). Assume first that $f^{-1}(p) = \{x\} \cup \{y\}$, that $\deg(f) = 1$ at x, and that $\deg(f) = -1$ at y. In $S^n \times I$ choose an embedded arc A connecting x and y and meeting $S^n - \{0\}$ transversally at x and y

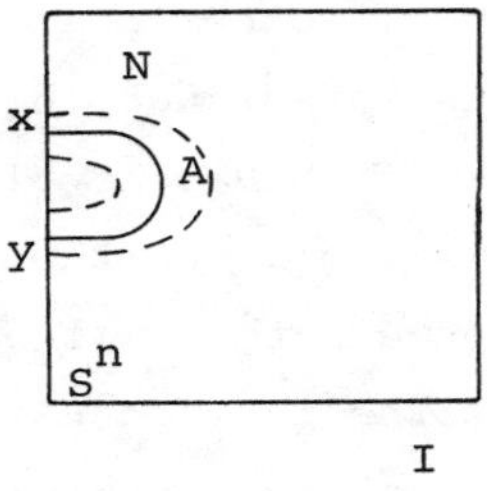

The disks around x and y (i.e., $f^{-1}(D^n)$) extend to a tubular neighborhood N of A. In fact, $N \cong e^n \times I$

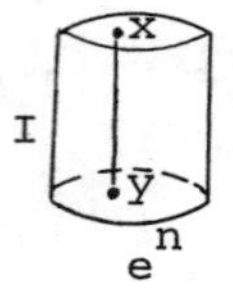

$\partial e^n = S^{n-1}$

Note that f is defined on $e^n \times \{0\}$ and $e^n \times \{1\}$ and, since the local degree of f at x and y has <u>opposite</u> signs, the degrees of f on $S^{n-1} \times \{0\}$ and $S^{n-1} \times \{1\}$ are the <u>same</u>. By the induction assumption f extends to a map $F: S^{n-1} \times I \to \partial D^n$. We may extend F to a map $F: e^n \times I \to D^n$ (this is a little exercise). Thus f extends to a map $\tilde{f}: S^n \times \{0\} \cup N \to S^n$ such that $\tilde{f}: \partial N \to \partial D^n \subset S^n$. Since S^n-int D^n is contractible we can extend $\tilde{f}$ to a map on the rest of $S^n \times I$. This map will have p as a regular value with preimage the arc A. The resulting map on $S^n \times \{1\}$ will miss p, and hence is homotopic to a constant.

If $f^{-1}(p)$ has more than 2 points and $n \geq 2$, then this argument allows us to deform f by a homotopy to "cancel" two of the points with opposite local degree. Continuing in this manner we can deform f until the preimage of p is empty if the deg(f) is 0. Once we have accomplished this the result is clearly homotopic to a constant map.

If n = 1, then we must take slightly more care. The point is that the arc A between two points of opposite sign must be disjoint from $x_j \times I$ for all other points $x_j \in f^{-1}(p)$. If $n \geq 2$, then general position ensures that this is always possible. If n = 1, then we must find a pair of points x and $y \in f^{-1}(p)$ of opposite local degree so that there are no other points of $f^{-1}(p)$ "in between them" (i.e., so that one of the two arcs in S^1 connecting them contains no other points of $f^{-1}(p)$). Since this is always possible if deg(f) = 0, we can still cancel a pair of points in this case.∎

There is a generalization of the Brouwer Theorem which we shall need.

<u>Theorem 3.2</u>: $H\colon \pi_n(\vee_i S^n) \to H_n(\vee_i S^n) \cong \oplus_i \mathbb{Z}$ <u>is an isomorphism for</u> $n > 1$.

The proof is analogous to the one given above. One takes $f\colon S^n \to \vee_i S_i^n$ and deforms it until it is regular with respect to a point $p_i \in S_i^n$ (p_i different from the point at which all the S_i^n are joined together). The cancelling argument shows that if the sum of the local degrees at $x_{ij} \in f^{-1}(p_i)$ is zero, then f is homotopic to a map missing p_i. (This requires that $n \geq 2$.) If $H(f) = 0$, then all the local degrees Σ_j (deg(f) at x_{ij}) $= 0$ for all i. Thus, if $H(f) = 0$, then f is homotopic to a map missing a point $p_i \in S_i$ for every i. This map is in turn homotopic to the constant map since $\vee_i(S^i - p_i)$ is contractible.∎

D. <u>The Hurewicz Theorem</u>.

<u>Statement</u>: <u>Let</u> X <u>be a CW complex. If</u> $\pi_k(X) = 0$ <u>for</u> $k < n$, <u>then</u>

(i) $\tilde{H}_k(X) = 0$ <u>for</u> $k < n$, <u>and</u>

(ii) $H\colon \pi_n(X) \to H_n(X)$ <u>is an isomorphism provided</u> $n > 1$.

<u>Proof</u>: We have already seen that if $\pi_k(X) = 0$ for $k < n$, then $\tilde{H}_k(X) = 0$ for $k < n$. It remains to show that

$$H\colon \pi_n(X) \xrightarrow{\cong} H_n(X).$$

<u>Step I</u>: Let us show that H is onto. Since $\pi_i(X) = 0$ for $i < n$ the argument given in the proof of the Whitehead theorem shows that there is a map $f\colon X \to X$ such that (i) f is homotopic to the identity, and (ii) $f(X^{(n-1)}) = x_0$. Let $a \in H_n(X)$, and let $\Sigma_{n\text{-cells}}\, a_\alpha(e_\alpha^n, \mathcal{O}_\alpha)$ be a cycle representative for a in $C_n(X)$. (Here, we choose arbitrarily an orientation $\mathcal{O}_\alpha$

for each e^n_α.) Clearly $f|e^n_\alpha: (e^n_\alpha, \partial e^n_\alpha) \to (X, x_0)$. Thus, $f|e^n_\alpha$ represents an element in $\pi_n(X, x_0)$. Since f is homotopic to the identity, $\alpha = f_*(\alpha) =$ class of $\Sigma_\alpha f(e^n_\alpha)$. Since each $f(e^n_\alpha)$ is represented by a sphere, this latter class is in the image of H.

We wish to show that $H: \pi_n(X) \to H_n(X)$ is injective, provided that $n > 1$ and $\pi_i(X) = 0$ for $i < n$. Let $f: S^n \to X$ be a map such that $f_*[S^n] = 0$ in $H_n(X)$. We can deform f until $f(S^n) \subset X^{(n)}$ and until f is transverse regular to a point p_i in the interior of each n-cell e^n_i. Let $\lambda_i = \Sigma_{x\epsilon f^{-1}(p_i}$ (local deg at x). The chain $\Sigma\ \lambda_i[e^n_i]$ is a cycle in $C_n(X)$ which represents $f_*[S^n] \in H_n(X)$. Since $f_*[S^n] = 0$ this cycle is a boundary; i.e. there exists μ_j such that $\Sigma\lambda_i[e^n_i] = \partial\Sigma\mu_j e^{n+1}_j$. Adding to $f: S^n \to X^{(n)}$ the linear combination $\Sigma\ \mu_j \partial e^{n+1}_j$ makes each $\lambda_i = 0$, and this does not change the homotopy class of $f: S^n \to X$. There is a map $\psi: X \to X$ such that (a) ψ is homotopic to id_X and (b) $\psi|X^{(n-1)}$ is constant. This means that $\psi|X^{(n)}$ factors through a wedge of spheres:

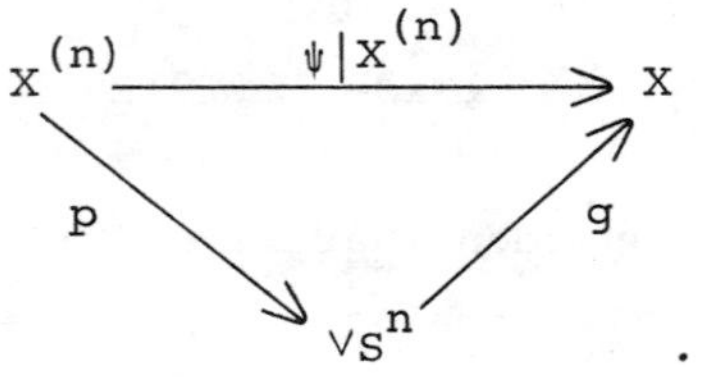

Since $f: S^n \to X^{(n)}$ has each $\lambda_i = 0$, the composition $p\circ f: S^n \to \vee S^n$ is homologous to zero. By 3.2 this means that $p\circ f: S^n \to \vee S^n$ is homotopic to zero. Hence $g\circ p\circ f = \psi\circ f$ is homotopically trivial. Since $\psi \simeq id_X$, this means that f itself is homotopically trivial.∎

<u>Note</u>: The Hurewicz theorem is valid for all spaces and not just CW complexes. To establish this one works with the cycles

themselves rather than with the spaces into which they are mapped. For a proof of the general result see Spanier's "Algebraic Topology".

E. Corollaries of the Hurewicz Theorem.

To begin with, there is (of course) a relative form. Let $A \subset X$ be a CW pair and assume that:

$$\pi_1(A) = 0$$

and

$$\pi_i(X,A) = 0, \quad i < n \quad (n \geq 2).$$

Then it follows that

$$H_i(X,A) = 0 \quad \text{for} \quad i < n$$

and the relative Hurewicz map

$$\pi_n(X,A) \xrightarrow[H]{} H_n(X,A)$$

is an isomorphism (the proof that $H_i(X,A) = 0$ for $i < n$ and that H is surjective follows from the proof of the Whitehead theorem, just as in the absolute case).

We now come to the corollaries of Hurewicz theorem.

Corollary 3.3: *If X, Y are simply connected CW complexes and $X \xrightarrow{f} Y$ induces an isomorphism on homology, then f is a homotopy equivalence.*

Proof: Let M_f be the mapping cylinder for f. Then $Y \hookrightarrow M_f$ is a deformation retract, and using this we make the identifications

$$\pi_i(Y) \cong \pi_i(M_f)$$

$$H_i(Y) \cong H_i(M_f).$$

The inclusion $X \hookrightarrow M_f$ gives exact sequences

$$\begin{array}{ccccccccc}
\cdots \to H_i(X) & \xrightarrow{f_*} & H_i(Y) & \to & H_i(M_f,X) & \to & H_{i-1}(X) & \xrightarrow{f_*} & \cdots \\
\uparrow & & \uparrow & & \uparrow & & \uparrow & & \\
\cdots \to \pi_i(X) & \xrightarrow{f_*} & \pi_i(Y) & \to & \pi_i(M_f,X) & \to & \pi_{i-1}(X) & \xrightarrow{f^*} & \cdots
\end{array} .$$

Using that f_* is an isomorphism on homology gives $H_i(M_f,X) = 0$ for $i \geq 0$. The relative Hurewicz theorem now gives

$$\pi_i(M_f,X) = 0 \quad \text{for } i \geq 0.$$

It follows then that M_f deformation retracts onto X (cf. the proof of the Whitehead theorem).∎

<u>Corollary 3.4</u>: <u>If</u> X <u>has the homotopy type of an</u> n-<u>dimensional CW complex and if</u> $\pi_i(X) = 0$ <u>for</u> $i \leq n$, <u>then</u> X <u>is contractible</u>.

<u>Proof</u>: Since $\pi_i(X) = 0$ for $i \leq n$, $H_i(X) = 0$ for $i \leq n$. On the other hand, $H_i(X) = 0$ for $i > n$ by dimension. Thus $\tilde{H}_*(X) = 0$ and so $\pi_*(X) = 0$ by the Hurewicz theorem.∎

<u>Corollary 3.5:</u> <u>If</u> X <u>has the homotopy type of an</u> n-<u>dimensional CW complex and if</u> $\pi_i(X) = 0$ <u>for</u> $i \leq n-1$, <u>then</u>

$$X \sim \vee S^n.$$

In particular, if $H_n(X) \cong \mathbb{Z}$, X *has the homotopy type of* S^n.

(Thus a simply-connected homology sphere is a homotopy sphere.)

Proof: This follows from the proof of the Hurewicz theorem in the following way: we may assume that $X = X^{(n)}$ and that we have

$$X \xrightarrow{g} \vee S^n \xrightarrow{f} X.$$

It follows that $H_n(\vee S^n) \xrightarrow{f_*} H_n(X)$ is onto and we may choose a basis for $H_n(\vee S^n) \cong \oplus H_n(S^n)$ (little exercise) and $H_n(X)$ such that f_* is given by a matrix

$$\underbrace{\left(\begin{array}{ccc|c} a_1 & & 0 & \\ & \ddots & & 0 \\ 0 & & a_b & \end{array}\right)}_{c} \Bigg\} b$$

($b = \text{rank } H_n(X)$; $c = \text{rank } H_n(\vee S^n)$, $a_i = \pm 1 \in \mathbb{Z}$).

Since $\pi_n(\vee S^n) \cong H_n(\vee S^n)$, we may find a map h

$$\begin{array}{ccc} \underbrace{S^n \vee \ldots \vee S^n}_{b} & & \\ {\scriptstyle h}\downarrow & \searrow {\scriptstyle j} & \\ \vee S^n & \xrightarrow[f]{} & X \end{array}$$

such that h_* has matrix

$$\underbrace{\left(\begin{array}{ccc} a_1 & & 0 \\ & \ddots & \\ 0 & & a_b \\ \hline & 0 & \end{array}\right)}_{b} \Bigg\} c$$

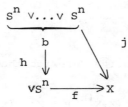

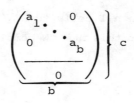

using the obvious basis for $H_n(\underbrace{S^n \vee \ldots \vee S^n}_{b})$.

It follows that j_* is an isomorphism on homology. ∎

Corollary 3.6: *If X is simply connected and $H_i(X) = 0$ for $i > n$, then $X \sim Y^{(n+1)}$; i.e., X has the homotopy type of an (n+1)-complex. If, in addition, $H_n(X)$ is free, then $X \sim Y^{(n)}$.*

Proof: We have from the CW homology theorem

$$H_i(X^{(n)}) \xrightarrow{\cong} H_i(X) \qquad i \neq n$$

$$H_n(X^{(n)}) \longrightarrow H_n(X) \longrightarrow 0$$

Using the exact homology sequence

$$H_i(X^n) \longrightarrow H_i(X) \longrightarrow H_i(X,X^{(n)}) \longrightarrow H_{i-1}(X^{(n)}) \longrightarrow H_{i-1}(X)$$

it follows that

$$H_i(X,X^{(n)}) = 0 \quad \text{for} \quad i \neq n+1$$

and that

$$0 \longrightarrow H_{n+1}(X,X^{(n)}) \longrightarrow H_n(X^{(n)}) \xrightarrow{i_*} H_n(X) \longrightarrow 0$$

is exact.

Using the relative Hurewicz theorem, we see that

$$\pi_{n+1}(X,X^{(n)}) \cong H_{n+1}(X,X^{(n)}).$$

Thus we may attach n+1 cells to $X^{(n)}$ to exactly kill the kernel of i_*. For this new CW complex $Y^{(n+1)}$,

$$H_i(Y^{(n+1)}) \cong H_i(X)$$

by construction for all i. Hence, $Y^{(n+1)}$ and X are homotopy equivalent. The other statement is an exercise. ■

Corollary 3.7: Let X be a simply-connected topological n-manifold. Then X has the homotopy type of an n-dimensional CW complex Y.

Remark: It is pretty clear that X has the homotopy type of a simplicial complex (c.f. exercise (13)). Thus, in the homotopy category X can be triangulated. If X is smooth then X can be smoothly triangulated (see Cairns-Hirsch, Annals of Math 41 (1940) pp. 796-808.) If X is a topological manifold, then it can usually be triangulated in a homogeneous fashion (see Kirby-Siebenmann, Annals of Math Studies No. 88). It is still possible that all topological manifolds can be triangulated in the sense that they all might be homeomorphic to simplicial complexes. (Siebenmann, pp. 77-84, Topology of Manifolds, ed. by J.C. Cantrell and C.H. Edwards, 1970.)

F. Homotopy theory of a Fibration.

Let $\pi: E \to B$ be a fibration (i.e., π has the homotopy lifting property). Let $\gamma: I \to B$ be a path from b_0 to b_1. We have a commutative diagram

$$\begin{array}{ccccc} \pi^{-1}(b_0) \times \{0\} & & \hookrightarrow & & E \\ \cap\downarrow & & & & \downarrow \pi \\ \pi^{-1}(b_0) \times I & \xrightarrow{p_2} & I & \xrightarrow{\gamma} & B \end{array} .$$

Applying homotopy lifting gives a map $\pi^{-1}(b_0) \times I \to E$ covering

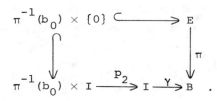

$\gamma \circ p_2$. It is easy to show that $\pi^{-1}(b_0) \times \{1\} \to \pi^{-1}(b_1)$ is a homotopy equivalence. Thus, we see that if B is path connected, then all fibers are homotopy equivalent.

If γ is a loop based at b_0, then the resulting homotopy equivalence $\pi^{-1}(b_0) \times \{1\} \to \pi^{-1}(b_0)$ is a homotopy automorphism of $\pi^{-1}(b_0)$. Its homotopy class depends only on the class of γ in $\pi_1(B,b_0)$. Thus, we have a representation $\pi_1(B,b_0) \to \mathrm{Auto}(\pi^{-1}(b_0))$ where $\mathrm{Auto}(\pi^{-1}(b_0))$ is the group of homotopy classes of homotopy equivalences of $\pi^{-1}(b_0)$. This representation is the _action of_ $\pi_1(B,b_0)$ _on the fiber_. There are induced actions on the homology and cohomology of the fiber.

Theorem 3.8: _Let_ $\pi\colon E \to B$ _be a fibration, and let_ $F = \pi^{-1}(b_0)$. _There is an exact sequence_:

$$\longrightarrow \pi_{n+1}(B,b_0) \xrightarrow{\partial} \pi_n(F,e_0) \xrightarrow{i_*} \pi_n(E,e_0) \xrightarrow{\pi_*} \pi_n(B,b_0) \longrightarrow$$

where $i\colon F \hookrightarrow E$ _is the inclusion_.

Proof: We have the long exact sequence of the pair $F \hookrightarrow E$.

$$\cdots \longrightarrow \pi_{n+1}(E,F) \xrightarrow{\partial} \pi_n(F) \longrightarrow \pi_n(E) \longrightarrow \pi_n(E,F) \longrightarrow \cdots \quad .$$

Also, we have $\pi_*\colon \pi_{n+1}(E,F) \to \pi_{n+1}(B,b_0)$. We claim that this map is an isomorphism. Once we know this the exact sequence as claimed is derived immediately from the above exact sequence of the pair.

We define an inverse to $\pi_*\colon \pi_{n+1}(E,F) \to \pi_{n+1}(B,b_0)$. Given $f\colon (I^n,\partial I^n) \to (B,b_0)$ there is a lifting

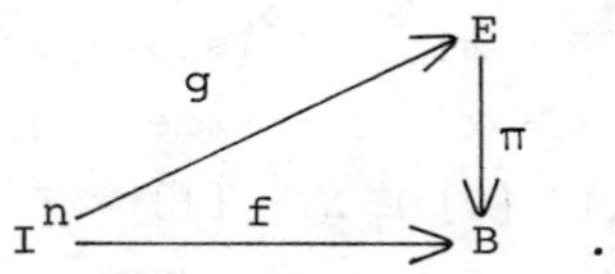

The reason is that since I^n is contractible, the map f is homotopic to the constant map $I^n \to b_0$. (The homotopy must in general deform ∂I^n off of b_0.) By the homotopy lifting property, any map, homotopic to a map which lifts, itself lifts. Clearly, $g: (I^n, \partial I^n) \to (E,F)$ determines an element in $\pi_n(E,F)$ which projects via π_* to f. This proves that π_* is onto.

To prove that it is one-to-one we show that if $g_0, g_1: (I^n, \partial I^n) \to (E,F)$ are such that $\pi \circ g_0$ is homotopic to $\pi \circ g_1$ as maps $(I^n, \partial I^n) \to (B, b_0)$, then $g_0 \simeq g_1$. By homotopy lifting we can assume that $\pi \circ g_0 = \pi \circ g_1$. The result now follows from the next lemma by letting $X = I^{n-1}$.

<u>Lemma 3.9</u>: <u>Given</u> $H: X \times I \to B$ <u>and two liftings</u> $\tilde{H}_1$ <u>and</u> $\tilde{H}_2: X \times I \to E$ <u>which agree in</u> $X \times \{0\}$, <u>then there is a homotopy of liftings of</u> H <u>connecting</u> $\tilde{H}_1$ <u>and</u> $\tilde{H}_2$ <u>all of which agree in</u> $X \times \{0\}$.

<u>Proof</u>: Let $\tilde{h} = \tilde{H}|X \times \{0\}$. We have a commutative diagram

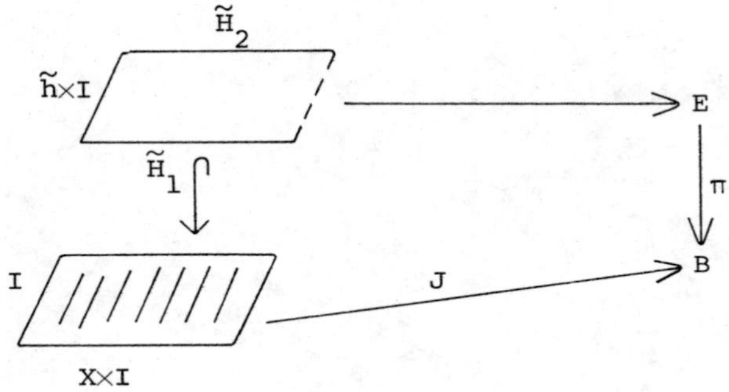

where the map $J: X \times I \times I \to B$ is projection onto $X \times I$ followed by H. Since $X \times I \times I$ deforms onto $X \times \{0\} \times I \cup X \times I \times \{0\} \cup X \times \{1\} \times I$ we can extend $\tilde{H}_1 \cup \tilde{h} \times I \cup \tilde{H}_2$ to a lifting of J. ∎

G. Applications of the exact homotopy sequence.

(i) $\pi_2(\mathbb{C}P^n) \cong \mathbb{Z}$ and $\pi_i(\mathbb{C}P^n) = \pi_i(S^{2n+1})$ for $i \neq 2$. This follows immediately from the homotopy long exact sequence of the Hopf fibration:

$$S^1 \longrightarrow S^{2n+1} \longrightarrow \mathbb{C}P^n$$

and the fact that $\pi_i(S^1) = 0$ for $i > 1$. To calculate the higher homotopy groups of S^1 recall that $\mathbb{R}^1 \xrightarrow{\exp} S^1$ is the universal cover. In general, the unique path lifting property of $\tilde{X} \to X$ implies that $\pi_i(\tilde{X}) \cong \pi_i(X)$ if $i > 1$ if $\tilde{X}$ is a covering space of X. Since $\mathbb{R}^1$ is contractible $\pi_i(S^1) \cong \pi_i(\mathbb{R}^1) = 0$ for $i > 1$.

(ii) $\pi_3(S^2) \cong \mathbb{Z}$.

This is the special case of $n = 1$ in the above example.

(iii) Let ΩB be the loop space on B. It is the fiber of $\mathcal{P}B \to B$ where $\mathcal{P}B$ is the path space of B. Since $\mathcal{P}B$ is contractible, this gives $\pi_{i-1}(\Omega B) \cong \pi_i(B)$.

4. Spectral Sequence of a Fibration

We begin with a fibration

$$\begin{array}{ccc} F & \longrightarrow & E \\ & & \big\downarrow \pi \\ & & B \end{array} .$$

In §3 we saw that the homotopy groups of F,E,B are related by an exact sequence.

$$\cdots \longrightarrow \pi_n(F) \longrightarrow \pi_n(E) \longrightarrow \pi_n(B) \xrightarrow{\partial} \pi_{n-1}(F) \longrightarrow \cdots$$

Our goal now is to understand how the cohomology of F, E, B is related. It is to be expected that the relationship will be somewhat complicated, because even in the case of a product $E = F \times B$, the Kunneth theorem giving $H^*(E)$ in terms of $H^*(F)$, $H^*(B)$ is more complicated than the simple formula

$$\pi_n(F \times B) \cong \pi_n(F) \oplus \pi_n(B)$$

for the homotopy groups.

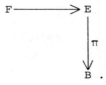

The total space of a fibration over a CW complex is filtered by the increasing sequence of subspaces lying over the various skeleta of the base. To calculate the cohomology of a pair of spaces one has a long exact sequence relating the cohomology of the subspace, the cohomology of the pair, and the cohomology of the total space. For a more general filtration there is a much more complicated algebraic formalism--a spectral sequence--which relates the cohomology of the successive pairs in the filtration and the cohomology of the space.

Suppose that B is a connected CW complex with p-skeleton $B^{(p)}$, and let $E^{(p)} = \pi^{-1}(B^{(p)})$. Let $b_0 \in B$ be the base point and $F = \pi^{-1}(b_0)$.

<u>Proposition 4.1</u>: <u>If</u> B <u>is path connected and if</u> $\pi_1(B,b_0)$ <u>acts trivially on</u> $H^*(F)$, <u>then there are isomorphisms</u>:

$$H^n(E^{(p)},E^{(p-1)}) \cong \prod_{\substack{\{p\text{-cells} \\ \text{in } B\}}} H^n(\pi^{-1}(e^p),\pi^{-1}(\partial e^p))$$
$$\cong C^p(B;H^{n-p}(F)).$$

<u>Proof</u>: Consider $\coprod_{\{p\text{-cells in } B\}} e^p \xrightarrow{\varphi} B$ to be the map of all p-cells into B. Form $\varphi^*E \to \coprod e^p$ the induced fibration. Since $(\coprod e^p, \coprod \partial e^p) \to (B^{(p)},B^{(p-1)})$ is a relative homeomorphism, so is $(\varphi^*E,\varphi^*E|\coprod\partial e^p) \to (E^{(p)},E^{(p-1)})$. Thus, by excision,

$$H^*(E^{(p)},E^{(p-1)}) \cong H^*(\varphi^*E,\varphi^*E|\ \partial e^p)$$

$$= \prod_{p\text{-cells}} H^*(\varphi^*E|e^p,\varphi^*E|\partial e^p).$$

Consider a fibration $E^1 \xrightarrow{\pi} e^p$. Let $F_0 = \pi^{-1}(0)$. We have a diagram

$$\begin{array}{ccc} F_0 \times \{0\} & \hookrightarrow & E^1 \\ \cap\downarrow & & \downarrow\pi \\ F_0 \times e^p & \xrightarrow{p_2} & e^p \end{array} .$$

Since e^p is contractible, the map p_2 lifts to a map $\widetilde{p}: F_0 \times e^p \to E^1$ extending the inclusion of F_0 into E'. This gives a fiberwise map $\widetilde{p}: (F_0 \times e^p, F_0 \times \partial e^p) \to (E^1, E^1|\partial e^p)$. Comparing the homotopy long exact sequences we see that $\widetilde{p}$ induces an isomorphism on the relative homotopy groups. Thus it is a homotopy equivalence of pairs.

Thus, in our case we have

$$H^*(\varphi^*E|e^p, \varphi^*E|\partial e^p) \cong H^*(F_0) \otimes H^*(e^p, \partial e^p).$$

If $\pi_1(B, b_0)$ acts trivially on $H^*(F)$, then choosing a path from b_0 to $0 \in e^p \subset B^{(p)}$ gives an identification of $H^*(F)$ with $H^*(F_0)$ which is independent of the path. Thus, we can identify $H^*(F_0) \otimes H^*(e^p, \partial e^p)$ with $H^*(F) \otimes H^*(e^p, \partial e^p)$. This gives

$$H^*(E^{(p)}, E^{(p-1)}) \cong \prod_{p\text{-cells}} H^*(e^p, \partial e^p) \otimes H^*(F)$$
$$\cong C^p(B; H^*(f)).$$

Returning now to our proposition, we may think of $H^*(E^{(p)}, E^{(p-1)})$ as a first approximation to $H^*(E^{(p)})$, of $H^*(E^{(p)}, E^{(p-2)})$ as a second and somewhat better approximation, of $H^*(E^{(p)}, E^{(p-3)})$ as a third approximation, etc. It is pretty clear that the successive approximations are related by exact sequences in which the missing terms are given by the previous proposition. Thus we can expect to get a whole chain of exact sequences, each related to the previous one, and each one giving us better and better approximation to $H^*(E)$. The algebra involved is formalized in the following.

<u>Definition</u>: A (first quadrant) <u>spectral sequence</u> consists of abelian groups $E_r = \{E_r^{p,q}\}_{p,q\geq 0}$ $(r \geq 0)$ and maps $d_r: E_r^{p,q} \to E_r^{p+r,q-r+1}$ with $d_r^2 = 0$, such that the homology $H(E_r) = E_{r+1}$. In detail:

$$E_{r+1}^{p,q} = \frac{\ker d_r: E_r^{p,q} \to E_r^{p+r,q-r+1}}{\operatorname{im} d_r: E_r^{p-r,q+r-1} \to E_r^{p,q}}.$$

<u>Remark</u>: The most important thing about spectral sequences is to have in mind the picture! For each term E_r, we plot the first quadrant in the (p,q) plane and put in the group $E_2^{p,q}$ at the lattice point (p,q). Then the differentials map according to the picture

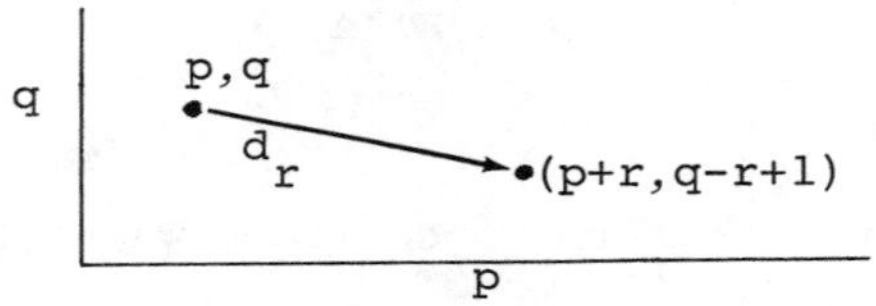

An element $\alpha \in E_r^{p,q}$ is said to <u>live to infinity</u> if $d_r\alpha = 0$ (and thus α defines an element in $E_{r+1}^{p,q}$), $d_{r+1}\alpha = 0,\ldots,$ etc. An element $\beta \in E_r^{p,q}$ is said to be <u>killed</u> if $d_r\beta = \ldots = d_{s-1}\beta = 0$ but $\beta = d_s\gamma$ for some $\gamma \in E^{p+s,q-s+1}$. The spectral sequence is said to degenerate at E_r if $E_r \cong E_{r+1} \cong \ldots$ etc.; it is said to be <u>degenerate</u> if $E_2 \cong E_3 \cong E_4 \cong \ldots$. Spectral sequences are a necessary and useful tool and should not be either over or underestimated. As with evaluation of calculus integrals, they are best learned by practice.

As an exercise in understanding pictures, we prove the

Lemma 4.2: *If* $N > p, q+1$ *then* $E_N^{p,q} \cong E_{N+1}^{p,q}$.

Proof: In the picture we find

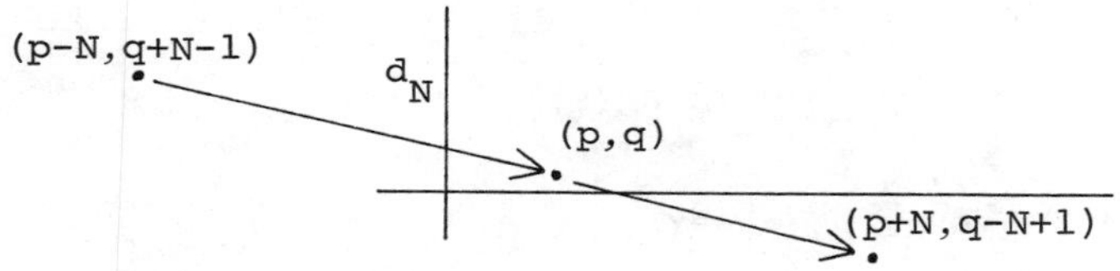

so that $d_N \equiv 0$ on $E_N^{p,q}$. ∎

Definition: We set $E_\infty^{p,q} = E_N^{p,q}$ for $N > p, q+1$ and $E_\infty^n = \oplus_{p+q=n} E_\infty^{p,q}$.

As another exercise in the definitions, we have the following:

Lemma 4.3: *Given a spectral sequence* $\{E_r^{p,q}\}$, *then*

$$0 \longrightarrow E_2^{1,0} \longrightarrow E_\infty^{1,0} \longrightarrow E_2^{0,1} \xrightarrow{d_2} E_2^{2,0} \longrightarrow E_\infty^2$$

is exact.

Proof: The picture is

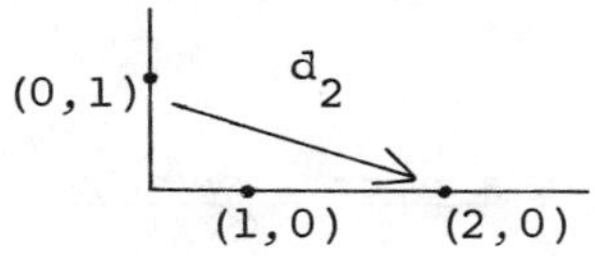

The main result concerning spectral sequences which interests us is the following.

Theorem 4.3 (Leray-Serre): Let $F \to E \to B$ be a fibration in which B is a connected CW complex with $\pi_1(B)$ acting trivially on the cohomology of the fiber. Then there is a spectral sequence $\{E_r^{p,q}\}$ in which

$$E_1^{p,q} = C^p(B, H^q(F)) \text{ and } d_1 = \text{coboundary map of } C^*(B, H^q(F));$$

$$E_2^{p,q} = H^p(B, H^q(F)); \text{ and}$$

$$E_N^{p,q} = \frac{\ker\{H^{p+q}(E) \to H^{p+q}(E^{(p-1)})\}}{\ker\{H^{p+q}(E) \to H^{p+q}(E^{(p)})\}}, \quad N > p, q+1.$$

Remarks: If we define

$$\mathcal{F}^p H^{p+q}(E) = \ker\{H^{p+q}(E) \longrightarrow H^{p+q}(E^{(p-1)})\}$$

then clearly

$$H^n(E) = \mathcal{F}^0 H^n(E) \supset \mathcal{F}^1 H^n(E) \supset \ldots \supset \mathcal{F}^n H^n(E) \supset \mathcal{F}^{n+1} H^n(E) = 0,$$

so that the groups $\mathcal{F}^p H^n(E)$ give a filtration on $H^n(E)$ whose associated graded module is $\oplus_{p+q=n} E_\infty^{p,q}$. This is usually written as

$$E_r^{p,q} \Longrightarrow H^{p+q}(E)$$

and one says that the spectral sequence converges (or abuts) to $H^*(E)$.

Proof: Let Sing*(E) be the singular cochains of E. We define a filtration

$$F^p(\mathrm{Sing}^*(E)) = \ker\{\mathrm{Sing}^*(E) \longrightarrow \mathrm{Sing}^*(E^{(p-1)})\}.$$

It is clear that $\delta F^p \subset F^p$. Thus, Sing*(E) is a cochain complex with a decreasing filtration preserved by δ. Under this circumstance it is a purely algebraic result that there is a spectral sequence which converges to the cohomology on Sing*(E). The terms are:

$$E_r^{p,q} = Z_r^{p,q}/B_r^{p,q}$$

$$Z_r^{p,q} = \{\text{cochains } \mu \text{ in } F^p(\mathrm{Sing}^{p+q}(E)) \text{ such that } \delta\mu \in F^{p+r}(\mathrm{Sing}^{p+q+1}(E))\}$$

$$B_r^{p,q} = \{Z_r^{p,q} \cap F^{p+1} + \delta F^{p-r+1}(\mathrm{Sing}^{p+q-1}(E)) \cap F^p\}.$$

It is a computational result that $H^*(E_r)$ (with respect to the differential d_r) is E_{r+1}. By definition $E_0^{p,q}$ is $F^p(\mathrm{Sing}^{p+q}(E))/F^{p+1}(\mathrm{Sing}^{p+q}(E)) \cong \mathrm{Sing}^{p+q}(E^{(p)},E^{(p-1)})$. The differential d_0 is the usual singular coboundary map. Thus $E_1^{p,q} = H^{p+q}(E^{(p)},E^{(p-1)})$. According to Proposition 4.1 $E_1^{p,q} \cong C^p(B;H^q(F))$. It is easy to see that under this identivication d_1 becomes $\delta_B \otimes 1$. Thus, $E_2^{p,q} = H^p(B;H^q(F))$. This completes the proof of the theorem.■

There is another version of the Leray-Serre spectral sequence for C^∞-manifolds. Let $\pi: E \to B$ be a C^∞ map between C^∞ manifolds which is locally (in B) a projection $F \times U \to U$. We define a filtration on the C^∞-differential forms of E. The condition that ω be in F^i is a pointwise condition which is required to hold for each point:

$$\omega^\ell \in F^i \Longleftrightarrow \langle \omega^\ell(p), \tau_1(p) \wedge \ldots \wedge \tau_\ell(p) \rangle = 0$$

when ever ℓ-i+1 of the tangent vectors $\tau_j(p)$ are in the kernel of $D\pi_p$.

<u>Explanation</u>: ω^ℓ is an ℓ-form and hence $\omega^\ell(p) \in \Lambda^\ell TE_p^*$. Thus, if $\tau_1(p), \ldots, \tau_\ell(p)$ are tangent vectors to E at p, then $\langle \omega^\ell(p), (\tau_1(p) \wedge \ldots \wedge \tau_\ell(p)) \rangle$ is a real number.

<u>Claim</u>: $E_0^{p,q} = \dfrac{F^p(\mathcal{E}^{p+q}(E))}{F^{p+1}(\mathcal{E}^{p+q}(E))}$ is identified with C^∞ p-forms on B with values in C^∞ q-forms on the fibers. (Here $\mathcal{E}^*(E)$ is the algebra of C^∞ forms on E.)

<u>Proof of Claim</u>: Let $\omega^{p+q} \in F^p(\mathcal{E}^{p+q}(E))$. Let $\tau_1(b), \ldots, \tau_p(b)$ be tangent vectors to B at $b = \pi(p)$. Let $\tilde{\tau}_i$ be a tangent vector to E at p which projects onto $\tau_i(b)$. Let $\tilde{\tau}_{p+1}, \ldots, \tilde{\tau}_{p+q}$ be tangent vectors to $\pi^{-1}(p) = F_p \subset E$. Then $\langle \omega^{p+q}(p), \tilde{\tau}_1 \wedge \ldots \wedge \tilde{\tau}_{p+q} \rangle$ is independent of the liftings $\tilde{\tau}_1, \ldots, \tilde{\tau}_p$ of the tangent vectors in B. This defines $F^p(\mathcal{E}^{p+q}) \to$ (p-forms on B with values in q-forms on the fibers). Clearly, this map factors through $F^p(\mathcal{E}^{p+q})/F^{p+1}(\mathcal{E}^{p+q})$. One checks easily that it induces an isomorphism

$$F^p(\mathcal{E}^{p+q})/F^{p+1}(\mathcal{E}^{p+q}) \cong \text{(p-forms on B with values in q-forms on the fibers).}$$

The differential d_0 becomes differentiation in the fiber direction under this identification. Hence

$$E_1^{p,q} = \mathcal{E}^p(B;H^q(F));$$

i.e., $E_1^{p,q}$ is p-forms on B with coefficients in the vector bundle of H^q along the fibers. One then shows that d_1 becomes $d_B \otimes 1$; i.e., differentiation in the base direction. Consequently,

$$E_2^{p,q} = H^p(B;H^q(F)) \quad \text{(real coefficients)}.$$

This is the C^∞ version of the Leray-Serre spectral sequence.

The last description we give of this spectral sequence is the one originally given by Serre (Annals of Math, Vol. 54).

One considers the singular cubical cochains Cube*(E). These again calculate the usual singular cohomology. We say that $\alpha \in \text{Cube}^n(E)$ is in filtration level i, $\alpha \in F^i$, if α vanishes when evaluated on any cubical chain $\varphi\colon I^n \to E$ which has the property that $\pi\circ\varphi\colon I^n \to B$ factors through the projection $I^n \to I^{i-1}$, $(t_1,\ldots,t_n) \to (t_1,\ldots,t_{i-1})$. This time $E_1^{p,q} = Q^p(B) \otimes H^q(F)$ where $Q^p(B)$ are the non-degenerate cubical cochains; i.e. the cubical cochains which annihilate all degenerate cubical chains ($\varphi\colon I^p \to B$ which factor through $I^p \to I^{p-1}$, $(t_1,\ldots,t_p) \to (t_1,\ldots,t_{p-1})$). Since the non-degenerate cubical cochains also calculate singular cohomology one finds that $E_2^{p,q} = H^p(B;H^q(F))$ ($\mathbb{Z}$-coefficients).

<u>Remarks</u>: (i) There is also a homology spectral sequence for a fibration, $\{E^r_{p,q}\}$, with $E^2_{p,q} = H_p(B;H_q(F))$. Over $\mathbb{Q}$ the two spectral sequences are dual in the obvious sense.

(ii) If $\pi_1(B,b_0)$ acts non-trivially on $H^*(F)$, then there is still a spectral sequence. In this case $E_2^{p,q}$ is $H^p(B;H^q(F))$ where the coefficients are twisted by the action.

(iii) There is a ring structure on the terms in the cohomology spectral sequence such that the differentials d_i

are <u>derivations</u>. In particular, if we use say $\mathbb{Q}$-coefficients, then:

$$E_2^{p,q} \cong H^p(B) \otimes H^q(F)$$

$$\cong E_2^{p,0} \otimes E_2^{0,q} \text{ and}$$

$$d_2 \cong d_2 \otimes 1 + (-1)^p\, 1 \otimes d_2 = (-1)^p\, 1 \otimes d_2$$

$$\text{since } d_2 = 0 \text{ on } E_2^{p,0}.$$

This is proved by using the usual cup product on cochains and being careful about the formulae; for more details, cf. Spanier, pp. 490-498. If one is willing to work over $\mathbb{R}$ (i.e., modulo torsion), then this multiplicative property is reasily seen using the differential form version of the spectral sequence discussed above. It should be remarked that the most sophisticated computations involving spectral sequences seem to use the ring structure and derivation property.

<u>Examples:</u>

(i) <u>complex projective space</u>. We shall compute the cohomology ring of $\mathbb{C}P^n = \mathbb{P}^n$ using the <u>Hopf fibration</u> $S^1 \to S^{2n+1} \to \mathbb{P}^n$.

Using $\mathbb{Z}$-coefficients, the E_2 term is $H^*(S^1) \otimes H^*(\mathbb{C}P^n)$ and $E_\infty \Rightarrow H^*(S^{2n+1})$. Since $H^q(S^1) = 0$ for $q > 1$, $E_2^{p,q} = 0$ for $q > 1$ and so $d_3 = d_4 = \ldots = 0$ and $E_3 \cong E_4 \cong \ldots \cong E_\infty$. Thus, all the "action" in the spectral sequence occurs at the E_2 level, and in particular $E_\infty^{p,q} = E_3^{p,q} = 0$ unless $p + q = 0, 2n+1$. The picture of E_2 is

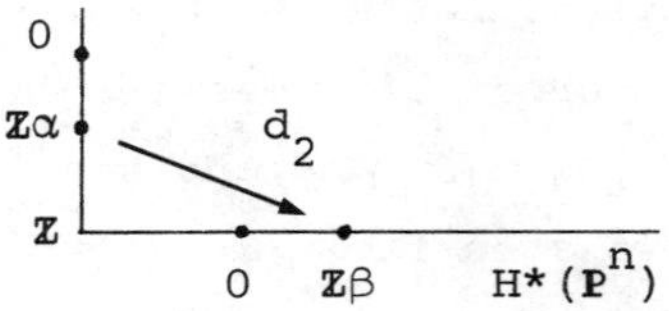

where $\alpha \in E_2^{0,1} \cong H^1(S^1)$ is a generator.

To begin with, $E_2^{1,0} = 0$ since $E_\infty^{1,0} \cong E_2^{1,0}$. Next the map

$$E_2^{0,1} \xrightarrow{d_2} E_2^{2,0}$$

is an isomorphism since $E_3^{p,q} = 0$ for $1 \leq p+q \leq 2$. Set $\beta = d_2\alpha$ so that $E_2^{2,1} \cong \mathbb{Z}\,(\alpha \otimes \beta)$. Then $d_2(\alpha \otimes \beta) = \beta^2 \in E_2^{4,0} \cong H^4(\mathbb{P}^n)$ since d_2 is a derivation. Continuing in this way we see that

$$H^{2q}(\mathbb{P}^n) = \mathbb{Z}\beta^q \qquad (q \leq n)$$

$$H^{2q+1}(\mathbb{P}^n) = 0$$

So that $H^*(\mathbb{P}^n) \cong \mathbb{Z}[\beta]/(\beta^{n+1})$; i.e., $H^*(\mathbb{P}^n)$ is a truncated polynomial algebra.

(ii) Rational cohomology of $K(\mathbb{Z},n)$.

We shall show in Section VI that there is a space, unique up to homotopy equivalence, which has only one non-zero homotopy group, that being π_n, and that group is isomorphic to $\mathbb{Z}$. Such a space is denoted $K(\mathbb{Z},n)$. We shall prove the fundamental result here that

$$H^*(K(\mathbb{Z},2k);\mathbb{Q}) \cong \mathbb{Q}[\alpha],$$

and

$$H^*(K(\mathbb{Z},2k+1);\mathbb{Q}) \cong \mathbb{Q}(\beta).$$

(The first is a polynomial algebra and the second is an exterior algebra.) The proof of these results is by induction on n. For $n = 1$, $K(\mathbb{Z},1) \cong S^1$ and the result is immediate. Consider the inductive step from $(2k-1)$ to $2k$. We study the path fibration

$$\begin{array}{ccc} K(\mathbb{Z},2k-1) & \longrightarrow & \mathcal{P} \\ & & \downarrow \\ & & K(\mathbb{Z},2k). \end{array}$$

Since $\mathcal{P}$ is contractible, the fiber is $K(\mathbb{Z},2k-1)$ and $E_\infty^{p,q} = 0$ if $(p,q) \neq (0,0)$. The E_2-term looks like:

0	
$H^*(K(\mathbb{Z},2k);\mathbb{Q})$	$(2k-1,*)$
0	
$H^*(K(\mathbb{Z},2k);\mathbb{Q})$	$(0,*)$.

Thus, $E_2 = E_3 = \ldots = E_{2k-1}$ and $E_{2k} = E_\infty$. First note that

$H^i(K(\mathbb{Z},2k);\mathbb{Q}) = \begin{cases} 0 & i < 2k \\ \mathbb{Q} & i = 2k \end{cases}$. (This is a consequence also of the Hurewicz theorem and the universal coefficient theorem). Let β be a generator of $H^{2k-1}(K(\mathbb{Z},2k-1);\mathbb{Q})$, and let $\alpha = d_{2k}(\beta)$. Then α is a generator of $H^{2k}(K(\mathbb{Z},2k);\mathbb{Q})$. Suppose inductively that we have shown that $\mathbb{Q}[\alpha] \to H^*(K(\mathbb{Z},2k);\mathbb{Q})$ is an isomorphism for all degrees $\leq t(2k)$. Then we have

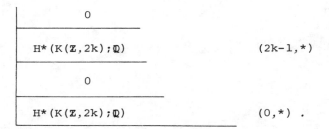

E_{2k}

	0							
β	0	$\beta \otimes \alpha$	0	$\beta \otimes \alpha^2$	0...0	$\beta \otimes \alpha^t$		
		0				0		
1	0...0	α	0...0	α^2	0...0...	α^t		$H^*(K(\mathbb{Z},2k);\mathbb{Q})$.

It results from the fact that $E_{2k+1}^{p,q} = 0$ for all $(p,q) \neq (0,0)$ that $H^i(K(\mathbb{Z},2k);\mathbb{Q}) = 0$ for $2kt < i < 2k(t+1)$ and that $H^{2k(t+1)}(K(\mathbb{Z},2k);\mathbb{Q})$ is isomorphic to $\mathbb{Q}$ and generated by $d_{2k}(\beta \otimes \alpha^t)$. By the derivation property $d_{2k}(\beta \otimes \alpha^t) = \alpha \otimes \alpha^t = \alpha^{t+1}$. This completes the induction step $(2k-1) \to 2k$.

For the inductive step $2k \to 2k+1$ we use the path fibration:

$$\begin{array}{ccc} K(\mathbb{Z},2k) & \longrightarrow & \mathcal{P} \\ & & \downarrow \\ & & K(\mathbb{Z},2k+1). \end{array}$$

Here E_2 is:

	⋮	
	0	
4k	α^2	
	0	
	⋮	
	0	
2k	α	
	0	
	⋮	
	$\mathbb{Q}$	$H^*(K(\mathbb{Z},2k+1);\mathbb{Q})$.

An argument similar to the one in the other case shows that

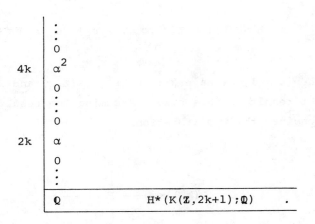

$$\begin{cases} E_2 = \ldots = E_{2k+1} \\ E_{2k+2} = \ldots = E_\infty \\ E_\infty^{p,q} = 0 \text{ for } p + q > 0 \end{cases} .$$

It follows that $E_{2k+1}^{0,2k} \xrightarrow{d_{2k+1}} E_{2k+1}^{2k+1,0}$ is an isomorphism, and so we set $\beta = d_{2k+1}\alpha$. By the derivation property, $d_{2k+1}(\alpha^2) = 2\alpha \otimes d_{2k+1}\alpha = 2\alpha \otimes \beta$, and so

$$E_{2k+1}^{0,4k} \xrightarrow{d_{2k+1}} E_{2k+1}^{2k+1,2k}$$

is an isomorphism (this is where we need $\mathbb{Q}$-coefficients). This gives that $H^{4k+2}(K(\mathbb{Z},2k+1),\mathbb{Q}) = 0$, and continuing on we obtain the desired result.

(iii) Grassmannians. Let $G(k,N)$ be the Grassmann manifold of k-planes in $\mathbb{C}^N$ and $G(k) = \lim_{N\to\infty} G(k,N)$ the infinite Grassmanian. Using $\mathbb{Z}$ coefficients we want to prove that:

$$H^p(G(k)) \cong \mathbb{Z}[c_1,c_2,\ldots,c_k], \quad c_j \in H^{2j}(G(k)).$$

When $k = 1$, we have $G(1) = \mathbb{C}P^\infty = K(\mathbb{Z},2)$ and we have proved the result in this case. Assuming the result for k-1, we consider the two fibrations

$$\begin{array}{ccccc} & & S^{2k-1} & & \\ & & \downarrow & & \\ G(k-1) & \longleftarrow & F & \longleftarrow & S^\infty \\ & & \downarrow & & \\ & & G(k) & & \end{array}$$

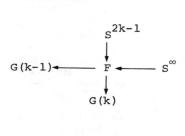

where $F = \{(v,\pi) \mid \pi$ is a k-plane in $\mathbb{C}^\infty$ and $v \in \pi$ is a unit vector$\}$. The map $F \to G(k-1)$ is given by sending (v,π) to the $(k-1)$ plane in π perpendicular to v. Clearly, $F \to G(k-1)$ is a fibration with fiber S^∞ which is contractible. Hence, $F \simeq G(k-1)$. The fibration $F \to G(k)$ is given by sending (v,π) to π. The fiber here is S^{2k-1}. Thus, we have a spectral sequence such that $E_2^{p,q} = H^p(G(k),H^q(S^{2k-1}))$ and E_∞ converges to $H^*(G(k-1))$. It must be the case that $E_2 = E_3 = \ldots = E_{2k-1}$ and that $E_{2k} = E_\infty$. Let c_k be $d_{2k}(\alpha)$ where α is the generator of $H^{2k-1}(S^{2k-1})$. From the spectral sequence we have exact sequences

$$H^{i+2k-1}(G(k-1)) \longrightarrow H^i(G(k)) \xrightarrow{\cup c_k} H^{i+2k}(G(k)) \longrightarrow H^{i+2k}(G(k-1)).$$

Since by induction $H^*(G(k-1))$ is a polynomial algebra generated by $c_1,\ldots,c_{k-1}$ and these classes lift to $H^*(G(k))$, it follows that $H^*(G(k)) \to H^*(G(k-1))$ is onto. Consequently, the above sequence becomes:

$$0 \longrightarrow H^i(G(k)) \xrightarrow{\cup c_k} H^{i+2k}(G(k)) \longrightarrow H^{i+2k}(G(k-1)) \longrightarrow 0 .$$

Using this one proves easily that $H^*(G(k))$ is a polynomial algebra on $c_1,\ldots,c_k$.

V. Obstruction Theory.

Recall, in the proof of the Whitehead theorem, we showed that if X is a CW complex and Y is a space with $\pi_i(Y) = 0$ for all $i \geq 0$, then any map $f: X \to Y$ is homotopic to the base point map $X \to y_0 \in Y$. The proof was by induction over the skeleta of X. Obstruction theory is a generalization of this technique to the case when the homotopy groups of Y are not necessarily zero. It does not give a complete understanding of when a map $f: X \to Y$ is homotopic to a constant but it gives some insight

Obstruction theory not only deals with the uniqueness question "When are two maps $f_1, f_2: X \to Y$ homotopic?", it also deals with the existence question of constructing maps from X to Y. In both cases we work inductively over the skeleta. In the first, we suppose that we have a homotopy h from $f_1|X^{(n)}$ to $f_2|X^{(n)}$ ($X^{(n)}$ = n-skeleton of the CW complex X):

$$h: X^{(n)} \times I \longrightarrow Y$$

and we ask if it extends to $X^{(n+1)} \times I \to Y$. In the second we suppose that we have a map $f: X^{(n)} \to Y$ and we ask if it extends to a map $\bar{f}: X^{(n+1)} \to Y$.

We will deal with the 2nd case first.

Lemma 5.1: If Y is simply connected, then there is a natural bijection

$$\pi_n(Y) \longleftrightarrow [S^n, Y],$$

where $[S^n, Y]$ means the free homotopy classes of maps of S^n into Y (no base point conditions).

Proof: Of course there is always a function

$$\pi_n(Y) \longrightarrow [S^n, Y]$$

obtained by ignoring the base points. If $\pi_0(Y) = 0$, then it is onto. This follows by applying the homotopy extension property (h.e.p.) to $S^n \times \{0\} \cup * \times I \subset S^n \times I$ (* is the base point of S^n). If in addition $\pi_1(Y) = 0$ then the map is 1-1. This follows by applying h.e.p. to $(S^n \times \{0 \cup 1\} \cup * \times I) \times I$ in $(S^n \times I) \times I$. ■

Assume for the rest of this section that $\pi_1(Y) = 0$.

Let (X,A) be a CW pair. Denote by $X^{(n)} \cup A$ the union of A and all the cells of X - A of dimension $\leq n$. Suppose given $f_n: X^{(n)} \cup A \to Y$. We define the obstruction cochain $\tilde{\mathcal{O}}(f_n) \in C^{n+1}(X,A;\pi_n(Y))$ as follows: If e_α^{n+1} is an oriented (n+1)-cell of (X,A) then its attaching map $c_\alpha: S^n \to X^{(n)} \cup A$ composed with f gives $f \circ c_\alpha: S^n \to Y$, which determines an element of $\pi_n(Y)$. If we reverse the orientation on e_α^{n+1} (and hence on ∂e_α^{n+1}), then the resulting element in $\pi_n(Y)$ changes sign. Thus, there is a well-defined homomorphism $C_{n+1}(X,A) \to \pi_n(Y)$. We denote it by $\tilde{\mathcal{O}}(f_n)$ and call it the obstruction cochain.

5.2: Properties of $\tilde{\mathcal{O}}(f_n)$:

1) It is an invariant of the homotopy class of f_n.
2) It is 0 if, and only if, f_n extends to a map $f_{n+1}: X^{(n+1)} \cup A \to Y$.
3) It is a cocycle; i.e., $\delta\tilde{\mathcal{O}}(f_n): C_{n+2}(X,A) \to \pi_n(Y)$ is 0.
4) If $g_n: X^{(n)} \cup A \to Y$ agrees with f_n on $X^{(n-1)} \cup A$, then $\tilde{\mathcal{O}}(g_n) - \tilde{\mathcal{O}}(f_n)$ is a coboundary.
5) By varying the homotopy class of f_n, relative to

$X^{(n-1)} \cup A$, we can change $\tilde{\mathcal{O}}(f_n)$ by an arbitrary coboundary.

Proof: 1) and 2) are immediate from the definitions. Before proving 3), 4), and 5), we review a little of the terminology associated with cohomology and then prove a necessary lemma.

Let $\{C_*, \partial\}$ be a chain complex. Its homology is defined by

$$H_n(C) = \frac{\ker \partial_n}{\operatorname{Im} \partial_{n+1}} = \frac{\text{cycles}}{\text{boundaries}} .$$

If G is an arbitrary abelian group, then we define a cochain complex with coefficients in G by

$$\cdots \longleftarrow \operatorname{Hom}(C_n, G) \xleftarrow{\partial^* = \delta} \operatorname{Hom}(C_{n-1}, G) \longleftarrow \cdots .$$

Clearly, $\delta \circ \delta = 0$. We define $H^n(C;G)$ to be $\dfrac{\ker \delta_n}{\operatorname{Im} \delta_{n-1}} = \dfrac{\text{cocycles}}{\text{coboundary}}$.

From this description it is easy to see that if σ_n is an n-cochain and τ_{n+1} is an (n+1)-chain, then $\langle \delta\sigma_n, \tau_{n+1} \rangle = \langle \sigma_n, \partial\tau_{n+1} \rangle$. Thus, σ_n is a cocycle if,and only if,it annihilates all boundaries.

Lemma 5.3: Let (X,A) be a simply connected CW-pair. Let $f: S^{n+1} \to X^{(n+1)} \cup A$ be a map with $f_*[S^{n+1}]$ represented by the CW cycle $\Sigma_\alpha a_\alpha e_\alpha^{n+1}$. Then f is homotopic to a map f' with $(f')^{-1}(\text{int } e_\alpha^{n+1})$ equal to $|a_\alpha|$ disjoint open balls in S^{n+1} each mapped homeomorphically onto int e_α^{n+1}.

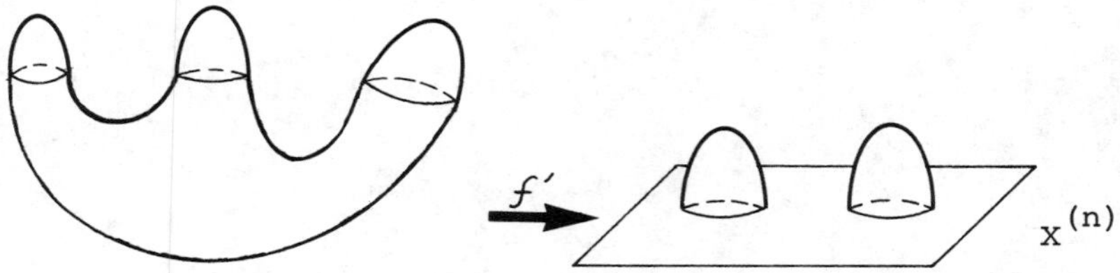

<u>Proof</u>: $f_*[S^{n+1}] = \Sigma_\alpha\, a_\alpha[e_\alpha^{n+1}, \partial e_\alpha^{n+1}] \in H_{n+1}(X^{(n+1)} \cup A, X^{(n)} \cup A)$. Since $\pi_1(X^{(n)} \cup A) = \pi_1(X) = 0$ $(n \geq 2)$ and $\pi_1(X^{(n+1)} \cup A, X^{(n)} \cup A) = 0$ for $i \leq n$, the Hurewicz theorem implies:

$$[f] = \Sigma_\alpha\, a_\alpha[e_\alpha^{n+1}, \partial e_\alpha^{n+1}] \in \pi_{n+1}(X^{(n+1)} \cup A, X^{(n)} \cup A).$$

Since $\partial[f] \in \pi_n(X^{(n)} \cup A)$ is zero, $\Sigma_\alpha\, a_\alpha[\partial e_\alpha^{n+1}] = 0$ in $\pi_n(X^{(n)} \cup A)$. Thus, $\coprod_\alpha a_\alpha \partial e_\alpha^{n+1}$ bounds a sphere minus $\Sigma_\alpha |a_\alpha|$ disks in $X^{(n)} \cup A$; call this $U \to X^{(n)} \cup A$. Clearly, $g: (\coprod_\alpha a_\alpha e_\alpha^{n+1}) \cup_\partial U \to X^{(n+1)} \cup A$ is an element of $\pi_{n+1}(X^{(n+1)} \cup A)$ whose image in $\pi_{n+1}(X^{(n+1)} \cup A, X^{(n)} \cup A)$ is the same as f. It follows that $[g] - [f]$ is in the image of $\pi_{n+1}(X^{(n)} \cup A) \to \pi_{n+1}(X^{(n+1)} \cup A)$. We change $g|U$ so that $[g] - [f] = 0$. This proves the lemma. ■

<u>Proof of 3</u>: $\delta(\tilde{\mathcal{O}}(f_n)) = 0$.

To prove this we need to show that $\langle \tilde{\mathcal{O}}(f_n), \partial\gamma_{n+2} \rangle = 0$ for all $(n+2)$-cells γ_{n+2}. The attaching map for γ, $\partial\gamma_{n+2}: S^{n+1} \to X^{(n+1)} \cup A$ is represented by $\Sigma_\alpha\, a_\alpha[e_\alpha^{n+1}, \partial e_\alpha^{n+1}]$ in $H_{n+1}(X^{(n+1)} \cup A, X^{(n)} \cup A)$. In the proof of the previous lemma we saw that $\Sigma_\alpha\, a_\alpha[\partial e_\alpha^{n+1}] = 0$ in $\pi_n(X^{(n)} \cup A)$. Thus,

$$\langle \tilde{\mathcal{O}}(f_n), \partial\gamma_{n+2} \rangle = \langle \tilde{\mathcal{O}}(f_n), \Sigma_\alpha\, a_\alpha[e_\alpha, \partial e_\alpha] \rangle$$

$$= \Sigma_\alpha\, a_\alpha \langle \widetilde{\mathcal{O}}(f_n), [e_\alpha, \partial e_\alpha] \rangle$$

$$= \Sigma_\alpha\, a_\alpha [f_n \circ \partial e_\alpha] \in \pi_n(Y)$$

$$= (f_n)_* \,\Sigma_\alpha\, a_\alpha [\partial e_\alpha]$$

$$= (f_n)_*(0) = 0.$$

<u>Proof of 4</u>: For each n-cell e_α^n in X - A, $f_n|\partial e_\alpha^n = g_n|\partial e_\alpha^n$. Thus, we can form the difference element "$f_n - g_n$": $S^n \to Y$.

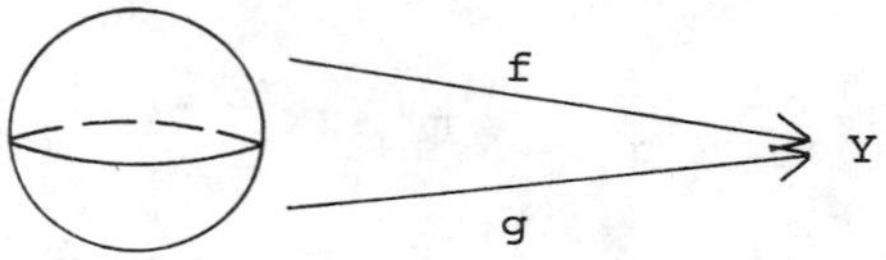

This defines a cochain "$f_n - g_n$": $C_n(X,A) \to \pi_n(Y)$.

<u>Claim:</u> $\delta(\text{"}f_n - g_n\text{"}) = \widetilde{\mathcal{O}}(f_n) - \widetilde{\mathcal{O}}(g_n)$.

<u>Proof</u>: Let e_α^{n+1} be an (n+1)-cell of (X,A) with $\partial e_\alpha^{n+1} = \Sigma\, a_{\alpha\beta} e_\beta^n$ in $C_n(X,A)$. As we have seen, ∂e_α^{n+1}: $S^n \to X^{(n)} \cup A$ and $\Sigma\, a_{\alpha\beta}[e_\beta^n]$ are equal as elements in $\pi_n(X^{(n)} \cup A, X^{(n-1)} \cup A)$. Deform ∂e_α^{n+1} until it is as in Lemma 5.3. Then f_n and g_n agree when restricted to $S^n - \Sigma_\beta |a_{\alpha\beta}|$ cells. On each of the cells there difference is "$f_n - g_n$" applied to the corresponding n-cell of (X,A). Thus

$$\langle \widetilde{\mathcal{O}}(f_n), \partial e_\alpha^{n+1} \rangle - \langle \widetilde{\mathcal{O}}(g_n), \partial e_\alpha^{n+1} \rangle = f_n \circ \partial e_\alpha^{n+1} - g_n \circ \partial e_\alpha^{n+1}$$

$$= \Sigma\, a_{\alpha\beta} \langle \text{"}f_n - g_n\text{"}, e_\beta^n \rangle = \langle (\text{"}f_n - g_n\text{"}), \partial e_\alpha^{n+1} \rangle$$

$$= \langle \delta(\text{"}f_n - g_n\text{"}), e_\alpha^{n+1} \rangle.$$

5) <u>We can vary $\tilde{\mathcal{O}}(f_n)$ by an arbitrary coboundary by changing f_n relative to $X^{(n-1)}$</u> $\cup$ A.

Let e_0^n be an n-cell of (X,A) and let g be an element of $\pi_n(Y)$. We shall show how to change f_n to f'_n so that $\tilde{\mathcal{O}}(f_n) - \tilde{\mathcal{O}}(f'_n) = \delta(\mu)$ where $\langle\mu, e_0^n\rangle = g$ and $\langle\mu, e^n\rangle = 0$ for all other n-cells e^n. Choose a small ball $B^n \subset \text{int } e_0^n$. By deforming f_n by a homotopy we can suppose that $f_n(B^n) = y_0 \in Y$. Define f'_n to agree with f_n on $X^{(n)} - \text{int } B^n$, and define $f'_n: (B^n, \partial B^n) \to (Y, y_0)$ to represent $g \in \pi_n(Y, y_0)$. Clearly, "$f'_n - f_n$" $= \mu$. Hence by the previous result $\tilde{\mathcal{O}}(f'_n) - \tilde{\mathcal{O}}(f_n) = \delta\mu$.

Let us consolidate our gains to this point.

<u>Theorem 5.4</u>: <u>Given</u> $f_n : X^{(n)} \cup A \to Y$ <u>with</u> $\pi_1(Y) = 0$, <u>there is a cohomology class</u> $\mathcal{O}(f_n) \in H^{n+1}(X,A;\pi_n(Y))$ <u>constructed from the cocycle</u> $\tilde{\mathcal{O}}(f_n)$. <u>This class vanishes if and only if</u> $f_n|X^{(n-1)} \cup A$ <u>can be extended to a map</u> $f: X^{(n+1)} \cup A \to Y$. <u>Let</u> f <u>and</u> $g: X \to Y$ <u>be given and</u> H: $(X^{(n)} \cup A) \times I \to Y$ <u>be a homotopy between</u> $f|X^{(n)} \cup A$ <u>and</u> $g|X^{(n)} \cup A$. <u>The obstruction to extending the homotopy over</u> $(X^{(n+1)} \cup A) \times I$ <u>lies in</u> $H^{n+1}(X \times I, ((X \times \{0 \cup 1\}) \cup A \times I); \pi_n(Y))$.

By the suspension isomorphism, this group is

$$H^n(X,A;\pi_n(Y)).$$

Thus,

(5.5): <u>The obstructions to constructing a homotopy between two maps</u> $f: X \to Y$ <u>and</u> $g: X \to Y$, <u>given a fixed homotopy on</u> A, <u>lie in</u> $H^n(X,A;\pi_n(Y))$.

<u>Discussion of the obstruction classes</u>: As long as $H^i(X,\pi_i(Y))$

is 0 we have no obstructions to finding a homotopy between f and g both restricted to $X^{(i)}$.

Suppose that $H^n(X,\pi_n(Y))$ is the first non-zero group.

It is an exercise (using the theorem one step at a time over the skeleta $X^{(i)}$, $i < n$) to show that, given f and g: $X \to Y$, the first obstruction to finding a homotopy between them, $\mathcal{O} \in H^n(X,\pi_n(Y))$, is well defined; i.e., does not depend on the step by step homotopy constructed from $f|X^{(n-1)}$ to $g|X^{(n-1)}$ this exhibits a universal phenomenon: <u>the first obstruction lying in a non-zero group is always well defined</u>. The higher obstructions are <u>not</u> defined until we make choices, and then depend on these choices.

As an unproved example, take $\mathbb{C}P^2 \xrightarrow{\varphi} S^2$ where $\varphi|\mathbb{C}P^1 = *$ (= base point in S^2) and φ on the 4-cell is given by a non-zero element in $\pi_4(S^2)$; namely $S^4 \xrightarrow{\Sigma\eta} S^3 \xrightarrow{\eta} S^2$ where $\Sigma\eta$ is the suspension of the Hopf map η. (We will do exercises of this type at a later time.) Take the point homotopy of $\varphi|\mathbb{C}P^1$ to $*$. Then the obstruction to extending this to a homotopy of φ to $*$ is in $H^4(\mathbb{C}P^2,\pi_4(S^2)) \cong \pi_4(S^2)$ and is the non-zero element ($\pi_4(S^2) \cong \mathbb{Z}_2$). But φ is homotopic to $*$. (Notice that $H^4(\mathbb{C}P^2,\pi_4(S^2))$ is not the first non-zero obstruction group since $H^2(\mathbb{C}P^2,\pi_2(S^2))$ is non-zero.)

Thus obstruction theory is not the be all and end all of homotopy theory. It was once described by Sullivan as much like being in a labirynth with a weak miner's light attached to your forehead and being forced always to move forward. The light enables you to see if you may take your next step but is not strong enough to tell you which fork to take when you must make a decision. Also there is no quarantee that, if you choose one path that eventually is blocked, then all of them are blocked. On the other hand, if you were a miner would you care to be without your light? Likewise, no topologist would forsake obstruction theory.

There is another type of obstruction theory which we

shall need later on. This is for the problem of constructing a section of a fibration. Let $p: E \to B$ be a fibration with B path connected and with $F = p^{-1}(b_0)$. A section $\sigma: B \to E$ is a map for which $p \circ \sigma = Id_B$. We assume that B is a CW complex, that $\pi_1(B)$ acts trivially on F, and that $\pi_1(F) = 0$. The obstructions to constructing a section lie in $H^i(B;\pi_{i-1}(F))$. Given two sections the obstructions to constructing a homotopy of sections connecting them lie in $H^i(B;\pi_i(F))$.

Let us give the definition of the obstruction cocycle. Suppose $\sigma: B^{(n-1)} \to E$ is a section over the (n-1) skeleton. For each n-cell of B we have a trivialization of $p^{-1}(e^n) \to e^n$ as $e^n \times F$, as we saw in Section IV. To construct such a trivialization one must connect e^n to b by a path in B. Since $\pi_1(B)$ acts trivially on the fiber, the identification is well-defined, up to homotopy, independently of the path chosen. A section σ on $B^{(n-1)}$ induces a section of $p^{-1}(e^n) \to e^n$ over ∂e^n. Using the above product structure this gives $\tilde{\sigma}: \partial e^n \to e^n \times F$. Projecting onto the second factor yields an element in $\pi_{n-1}(F)$. The obstruction cocycle $\tilde{\mathcal{O}}_n(\sigma): C_n(B) \to \pi_{n-1}(F)$ assigns to e^n the resulting homotopy element.

Arguments similar to the ones above show that $\tilde{\mathcal{O}}_n(\sigma)$ is a cocycle and that one may vary $\tilde{\mathcal{O}}_n(\sigma)$ by an arbitrary coboundary by changing σ in $B^{(n-1)} - B^{(n-2)}$. Thus, the class $\mathcal{O}_n(\sigma) \in H^n(B;\pi_{n-1}(F))$ is the obstruction to extending $\sigma|B^{(n-2)}$ over $B^{(n)}$.

If (B,A) is a CW pair and we are given a section over $B^{(n)} \cup A$, then the obstruction to extending its restriction to $B^{(n-1)} \cup A$ over $B^{(n+1)} \cup A$ lies in $H^{n+1}(B,A;\pi_n(F))$.

Let us consider some examples of these obstruction classes.

Example A) The Euler class: If $E^n \to B$ is an n-dimensional vector bundle, then a nowhere zero section of E^n is the same as a section of the associated sphere bundle $S^{n-1}(E)$. The first obstruction to finding a section is in $H^n(B,\pi_{n-1}(S^{n-1})) \cong H^n(B;\mathbb{Z})$. It is called the Euler class of E^n and is an unstable characteristic class of the vector bundle. Some of the exercises deal with it in more detail.

Notice that this shows that if $\dim B < n$, then $E^n \to B$ always has a non-zero section, and thus as a vector bundle $E^n \cong E^{n-1} \oplus \epsilon^1$, where ϵ^1 is the trivial line bundle.

Example B) Eilenberg-MacLane spaces $K(\pi,n)$: Given $n \geq 2$ and π an abelian group, we will show that up to homotopy equivalence there is exactly one CW complex $K(\pi,n)$ such that $\pi_i(K(\pi,n)) = \begin{cases} 0 & i \neq n \\ \pi & i = n \end{cases}$.

Suppose that π is finitely presented

$$0 \longrightarrow F(r_1,\ldots,r_\ell) \longrightarrow F(\alpha_1,\ldots,\alpha_k) \longrightarrow \pi \longrightarrow 0$$

where $F(\ ,\ldots,\ )$ is the free abelian group generated by the symbols inside the parentheses. Form the space $(\vee_{i=1}^k S^n_{\alpha_i})$. Its nth homotopy group is $F(\alpha_1,\ldots,\alpha_k)$ (here we assume $n > 1$). Thus the elements r_i are represented by maps $\varphi_{r_j} : S^n \vee_{i=1}^k S^n_{\alpha_i}$. Form the (n+1)-complex $(\vee_{i=1}^k S^n_{\alpha_i}) \cup (\cup_{j=1}^\ell e^{n+1}_{r_j})$ where the attaching map for $e^{n+1}_{r_j}$ is φ_{r_j}. Call this CW complex $X^{(n+1)}$. Clearly $\pi_n(X^{(n+1)}) = \pi$ and $\pi_i(X^{(n+1)}) = 0$ if $i < n$. Choose a generating set $\{\gamma_s\}_{s\in S}$ for $\pi_{n+1}(X^{(n+1)})$ and use these elements to attach (n+2)-cells. Call the result $X^{(n+2)}$. One sees that:

$$\pi_i(X^{(n+2)}) = \begin{cases} 0 & i \leq n+1 \text{ and } i \neq n \\ \pi & i = n. \end{cases}$$

Continuing in this manner we construct

$$X^{(n)} \subset X^{(n+1)} \subset X^{(n+2)} \subset \ldots$$

so that $\pi_i(X^{(n+k)}) = 0$ for $i < n+k$ and $i \neq n$, whereas $\pi_n(X^{(n+k)}) = \pi$. Let X be the union of the $X^{(n+k)}$. It has the correct homotopy groups.

There is a similar construction even if π is not finitely presented.

If Y is a CW complex with the same homotopy groups as X, then Y is homotopy equivalent to X. To establish this we construct $f\colon X \to Y$ which induces an isomorphism on the homotopy groups. Begin with a map $f_n\colon \vee_{i=1}^{k} S^n_{\alpha_i} \to Y$ which induces the composition:

$$\pi_n(\vee S^n_{\alpha_i}) = F(\alpha_1,\ldots,\alpha_k) \longrightarrow \pi = \pi_n(Y).$$

This map extends over $X^{(n+1)}$ since

$$\varphi_{r_j}\colon S^n \longrightarrow \vee_{i=1}^{k} S^n_{\alpha_i}$$

represents an element in $\pi_n(\vee S^n_{\alpha_i})$ which is trivial in $\pi = \pi_n(Y)$. Any extension $f_{n+1}\colon X^{(n+1)} \to Y$ induces the identity

$$\pi_n(X^{(n+1)}) = \pi \xrightarrow{\text{id}} \pi = \pi_n(Y).$$

The obstructions to extending f_{n+1} over all of X lie in

$H^*(X:\pi_{*-1}(Y))$ for $* \geq n+2$. Hence, all obstructions vanish and f_{n+1} extends to $f: X \to Y$. Since $f_*: \pi_n(X) \to \pi_n(Y)$ is an isomorphism, f is a homotopy equivalence.

Remark: $K(\mathbb{Z},1) = S^1$ and $K(\mathbb{Z},2) = \mathbb{C}P^\infty$, but outside of these the $K(\pi,n)$ are usually not spaces encountered directly in geometry. In general, the price for having the π_i so simple is that the homology $H_i(K(\pi,n), \mathbb{Z})$ is now very complicated (this seems quite plausible from the construction, since we are adding cells in all dimensions according (roughly) to the pattern of $\pi_k(S^n)$ $(k > n)$).

Using the $K(\pi,n)$ we can construct a space with preassigned homotopy groups (take $X = \Pi K(\pi_i,n_i)$). Later we will show that any simply connected CW complex is homotopy equivalent to an iterated fibration of $K(\pi_i,n_i)$'s. For the moment we want to record one important fact about obstruction classes: the first possible non-zero class is well defined and natural. We establish this in the two contexts of extending a map and a section.

Theorem 5.6: Let (X,A) be a CW-pair, and let $f: A \to Y$ be given. Suppose $H^i(X,A;\pi_{i-1}(Y)) = 0$ and $H^{i-1}(X,A;\pi_{i-1}(Y)) = 0$ for $i \leq n$. Also suppose $\pi_1(Y) = 0$. The first obstruction to extending f over X, $\mathcal{O} \in H^{n+1}(X,A;\pi_n(Y))$ is well defined. It is natural with respect to maps $\varphi: (X',A') \to (X,A)$.

Proof: As we extend f over $X^{(i)} \cup A$ all obstructions lie in the zero group until we get to $X^{(n)} \cup A$. Let f_n and f'_n be two extensions over $X^{(n)} \cup A$. Since $H^i(X,A;\pi_i(Y)) = 0$ for $i \leq n-1$, $f'_n|(X^{(n)} \cup A)$ is homotopic relative to A to $f_n|(X^{(n)} \cup A)$. Thus, by 5.2, $\mathcal{O}(f_n) = \mathcal{O}(f'_n)$ in $H^{n+1}(X,A;\pi_n(Y)$. This proves that the primary obstruction is well defined.

Now suppose given $\varphi: (X',A') \to (X,A)$ where $H^*(X',A';\pi_{*-1}(Y))$

and $H^{*-1}(X',A';\pi_{*-1}(Y))$ are equal to zero for $* \leq n$. Then the first obstruction to extending $f\circ\varphi$: $A' \to Y$ lies in $H^{n+1}(X',A';\pi_*(Y))$. As above we see that it is well defined. Naturality means that $\varphi^*(\mathcal{O}(f)) = \mathcal{O}(f\circ\varphi)$ in $H^{n+1}(X',A';\pi_n(Y))$. Deform φ to a cellular map, φ', relative to A'. Let $f_n: X^{(n)} \cup A \to Y$ be an extension of $f: A \to Y$. Then $f_n\circ\varphi': X'^{(n)} \cup A \to Y$ is an extension of $f\circ\varphi$ on A'. We claim that the cocycle obstruction for $f_n\circ\varphi'$ is the pull back via φ' of the cocycle obstruction for f_n.

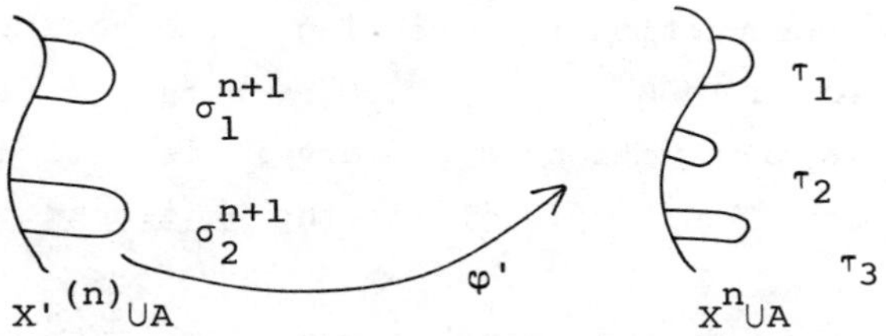

If $\varphi'_*(\sigma_1) = \Sigma_i\ \alpha_{1i}\tau_i$ in $C_{n+1}(X,A)$, then $\varphi'_*(\partial\sigma_1) = \Sigma_i\ \alpha_{1i}\partial\tau_i$ in $\pi_n(X^{(n)} \cup A)$. Thus

$$\begin{aligned}
\langle\mathcal{O}(f_n\circ\varphi'),\sigma_i\rangle &= f_n\circ\varphi'(\partial\sigma_i)\\
&= f_n(\Sigma_i\ \alpha_{1i}\partial\tau_i)\\
&= \Sigma_i\ \alpha_{1i}f_n(\partial\tau_i)\\
&= \Sigma_i\ \alpha_{1i}\langle\mathcal{O}(f_n),\tau_i\rangle\\
&= \langle\mathcal{O}(f_n),\Sigma_i\ \alpha_{1i}\tau_i\rangle\\
&= \langle\mathcal{O}(f_n),\varphi'_*(\sigma_1)\rangle\\
&= \langle(\varphi')^*\mathcal{O}(f_n),\sigma_1\rangle. \quad \blacksquare
\end{aligned}$$

Examples: 1) If $S^n \xrightarrow{f} Y$, then the obstruction to f being homotopic to zero is in $H^n(S^n;\pi_n(Y)) = \pi_n(Y)$. It is, of course, the homotopy class of f in $\pi_n(Y)$.

2) Let $g\colon (\mathbb{C}P^2)^{(3)} \to S^2$ be the identity map. The obstruction to extending g over $\mathbb{C}P^2$ is in $H^4(\mathbb{C}P^2;\pi_3(S^2)) = \pi_3(S^2)$. It is the Hopf map; i.e., the attaching map for the 4-cell of $\mathbb{C}P^2$. This is true in general. If $X' = X \cup_f e^n$, then the obstruction to extending $\mathrm{Id}\colon X \to X$ over X' is $[f] \in \pi_{n-1}(X)$.

3) Given $f\colon S^k \to Y$ and $g\colon S^\ell \to Y$, form $f \vee g\colon S^k \vee S^\ell \to Y$. The only obstruction to extending $f \vee g$ to a map $S^k \times S^\ell \to Y$ is an element in $H^{k+\ell}(S^k \times S^\ell, S^k \vee S^\ell;\pi_{k+\ell-1}(Y) \cong \pi_{k+\ell-1}(Y)$. Since this obstruction is primary it is well defined. The obstruction, denoted $[f,g]$, is the Whitehead product of f and g.

There is an analogous theorem for lifting maps in a fibration.

Theorem 5.7: Let $\pi\colon E \to X$ be a fibration with fiber F. Suppose that $\pi_1(X)$ acts trivially on F and that $\sigma\colon A \to E|A$ is a section of π over A. If $H^i(X,A;\pi_{i-1}(F)) = 0$ and $H^i(X,A;\pi_i(F)) = 0$ for $i \leq n$, then the first obstruction to extending σ on X lies in $H^{n+1}(X,A;\pi_n(F))$. It is well defined and natural.

The proof is analogous to that of 5.6 and is left to the reader.

VI. Eilenberg-MacLane Spaces, Cohomology, and Principal Fibrations

A. Relation of cohomology and Eilenberg-MacLane Spaces.

In the last section we saw that there was a natural transformation $[(X,A),(K(\pi,n),*)] \to H^n(X,A;\pi)$ which assigns to any map $f\colon (X,A) \to (K(\pi,n),*)$ the primary obstruction to deforming f to a constant map relative to A. (Here, $*$ is the base point of $K(\pi,n)$.) Actually, this should be viewed as an extension problem for the map $f \cup c \cup c\colon X \times \{0\} \cup A \times I \cup X \times \{1\} \to K(\pi,n)$ (where c denotes the constant map). Thus, the primary obstruction is well-defined and lies in

$$H^{n+1}(X \times I, X \times \{0\} \cup A \times I \cup X \times \{1\};\pi) \cong H^n(X,A;\pi).$$

By the Hurewicz theorem $H_n(K(\pi,n)) = \pi$ and $H_{n-1}(K(\pi,n)) = 0$. Thus, by the universal coefficient theorem, $H^n(K(\pi,n);\pi) = \mathrm{Hom}(\pi,\pi)$. Here, we are viewing $K(\pi,n)$ as a space together with an identification of $\pi_n(K(\pi,n))$ with π. Let $\iota \in H^n(K(\pi,n);\pi)$ be the class corresponding to the identity homomorphism. If $f\colon (X,A) \to (K(\pi,n),*)$, then we have $f^*\iota \in H_n(X,A;\pi)$. This defines a function

$$i\colon [(X,A),(K(\pi,n),*)] \longrightarrow H^n(X,A;\pi).$$

<u>Theorem 6.1</u>: $f^*\iota$ <u>is the primary obstruction to deforming</u> f

to a constant relative to A. The association $[f] \to f^*\iota$ is a bijection for all CW-pairs (X,A).

Proof: First we show that $f^*\iota$ is the first obstruction to deforming f to a constant map relative to A. By naturality it suffices to show that the obstruction for the identity map $K(\pi,n) \to K(\pi,n)$ is ι. This is immediate from the definitions.

Since $K(\pi,n)$ has only one non-zero homotopy group, if $f^*\iota = 0$ then the one and only obstruction to deforming f to a point, relative to A, vanishes. It follows that in that case f is homotopic, rel A, to a constant map. This shows that $\iota^{-1}(0)$ is exactly the trivial map. This does not yet suffice to prove that ι is 1-1 because we have not defined a group structure on $[(X,A),(K(\pi,n),*)]$; cf. below.

We show that ι is onto. Let $\alpha \in H^n(X,A;\pi)$ be a class and let $\tilde{\alpha}: C_n(X,A) \to \pi$ be a cocycle representative for α. Define $f_{\tilde{\alpha}}|X^{(n-1)} \cup A$ to be the constant map. Define $(f_{\tilde{\alpha}}|e^n): e^n \to K(\pi,n)$ to be a map representing $\tilde{\alpha}(e^n) \in \pi = \pi_n(K(\pi,n))$. This gives $f_{\tilde{\alpha}}|X^{(n)} \cup A$. Since $\tilde{\alpha}$ is a cocycle, if we compose $f_{\tilde{\alpha}}|X^{(n)} \cup A$ with the attaching map for an (n+1)-cell the result is homotopically trivial. Thus $f_{\tilde{\alpha}}|X^{(n)} \cup A$ extends over $X^{(n+1)} \cup A$. All other obstructions to extending $f_{\tilde{\alpha}}|X^{(n)} \cup A$ vanish. Let $f_{\tilde{\alpha}}: (X,A) \to K(\pi,n)$ be an extension of $f_{\tilde{\alpha}}|X^{(n)} \cup A$. Clearly, the relative cocycle which measures the primary obstruction to a homotopy from $f_{\tilde{\alpha}}$ to the constant map is exactly $\tilde{\alpha} \in \mathrm{Hom}(C_n(X,A),\pi)$. Thus, $f_{\tilde{\alpha}}^*\iota = \alpha$.

To see that ι is 1-1 we assume that f and g map (X,A) to $(K(\pi,n),*)$ and $f^*\iota = g^*\iota$. The primary (and only) obstruction to a homotopy, rel A, from f to g is in $H^n(X,A;\pi)$. It is easily identified with $f^*\iota - g^*\iota = 0$. Thus,

f is homotopic to g relative to A. ■

<u>Application</u>: There is a map $K(\pi,n) \times K(\pi,n) \xrightarrow{\mu} K(\pi,n)$ so that $\mu^*\iota = \iota \otimes 1 + 1 \otimes \iota \in H^n(K(\pi,n) \times K(\pi,n);\pi) \cong [H^n(K(\pi,n);\pi) \otimes H^0(K(\pi,n);\mathbf{Z})] \oplus [H^0(K(\pi,n);\mathbf{Z}) \otimes H^n(K(\pi,n);\pi)]$. It is well defined up to homotopy. It acts as a commutative, associative group multiplication up to homotopy. Thus, $K(\pi,n)$ is a <u>homotopy commutative, associative</u> H-<u>space</u>. This map μ induces an abelian group structure on $[(X,A);(K(\pi,n),*)]$. With this group structure ι becomes a group isomorphism. (Details of proof left as exercises.)

B. <u>Principal</u> $K(\pi,n)$-<u>fibrations</u>.

A map $p: E \to B$ is said to be a $K(\pi,n)$-fibration if it satisfies the homotopy lifting property and if all fibers, $p^{-1}(b)$, are spaces of type $K(\pi,n)$. If B is path connected, then $p: E \to B$ is a $K(\pi,n)$-fibration provided that $p^{-1}(b)$ is a space of type $K(\pi,n)$ for any $b \in B$.

A $K(\pi,n)$-fibration is said to be <u>principal</u> if the action of the fundamental group of the base on the fiber is trivial up to homotopy. For each loop γ in the base B based at b there is a self-homotopy equivalence $\gamma_*: p^{-1}(b) \to p^{-1}(b)$ well-defined up to homotopy. The fibration is principal if all these self-equivalences are homotopic to the identity.

<u>Lemma 6.2</u>: <u>Let</u> (X,A) <u>be a</u> CW <u>pair, let</u> $p: E \to X$ <u>be a principal</u> $K(\pi,n)$ <u>fibration, and let</u> $\sigma: A \to E$ <u>be a section of</u> E <u>over</u> A. <u>There is a unique obstruction</u> $\mathcal{O}(p,\sigma) \in H^{n+1}(A;\pi)$ <u>to extending</u> σ <u>over all of</u> X. <u>Given any class</u> $\mathcal{O} \in H^{n+1}(X,A;\pi)$ <u>it is realized as the obstruction</u> $\mathcal{O}(p,\sigma)$ <u>for some principal fibration</u> $p: E \to X$ <u>and some section</u> $\sigma: A \to E$.

<u>Proof</u>: Since $\pi_i(\text{fiber}) = 0$ for $i < n$, according to Theorem 5.7

the first obstruction to extending the section lies in $H^{n+1}(X,A;\pi)$. It is well defined and natural. Since all the higher homotopy groups of the fiber vanish $\mathcal{O}(p,\sigma)$ is the unique obstruction to extending the fiber over all of X. Given a class $\mathcal{O} \in H^{n+1}(X,A;\pi)$ there is a map

$$f_{\mathcal{O}}: (X,A) \longrightarrow (K(\pi,n+1),*)$$

so that $f^*_{\mathcal{O}}(\iota) = \mathcal{O}$. Over $K(\pi,n+1)$ we have the principal fibration

$$\begin{array}{ccc} K(\pi,n) \cong \Omega K(\pi,n+1) & \longrightarrow & \mathcal{P}K(\pi,n+1) \\ & & \downarrow \\ & & K(\pi,n+1) \end{array}$$

Since $K(\pi,n+1)$ is simply connected this fibration is principal. If we pull back the fibration by $f_{\mathcal{O}}$, then we get a principal $K(\pi,n+1)$ fibration over X. Any section over $*$, pulls back to a section over A. The obstruction to extending the section over $* \in K(\pi,n+1)$ to one over all of $K(\pi,n+1)$ is $\iota \in H^{n+1}(K(\pi,n+1),\pi)$. By naturality the obstruction to extending the induced section over A to one over all of X is $f^*_{\mathcal{O}}(\iota) = \mathcal{O}$. This proves all classes arise as obstructions. ∎

The results of this section are summarized in the following commutative diagram.

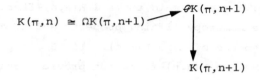

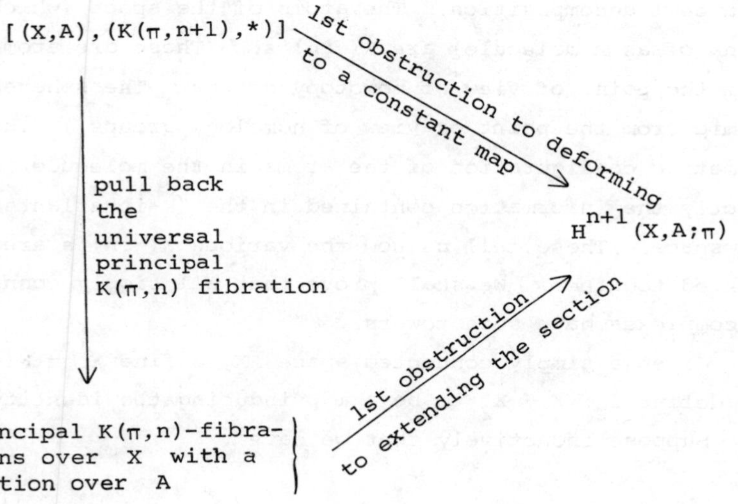

The map in the upper right hand corner is a bijection.

VII. Postnikov Towers and Rational Homotopy Theory

A Postnikov tower for a space is a decomposition dual to a cell decomposition. The atoms of the space (which we think of as a molecule) are $K(\pi,n)$'s. (These are atomic from the point of view of homotopy groups. The spheres are atomic from the point of view of homology groups.) The geometric configuration of the atoms in the molecule is exactly the information contained in the "k-invariants" of the space. These tell us how the various $K(\pi,n)$'s are twisted together. We shall prove that all simply connected CW complexes have such towers.

Given a simply connected space X, define $X_2 = K(\pi_2(X),2)$, and define $f_2\colon X \to X_2$ to be a map inducing the identity on π_2. Suppose inductively that we have

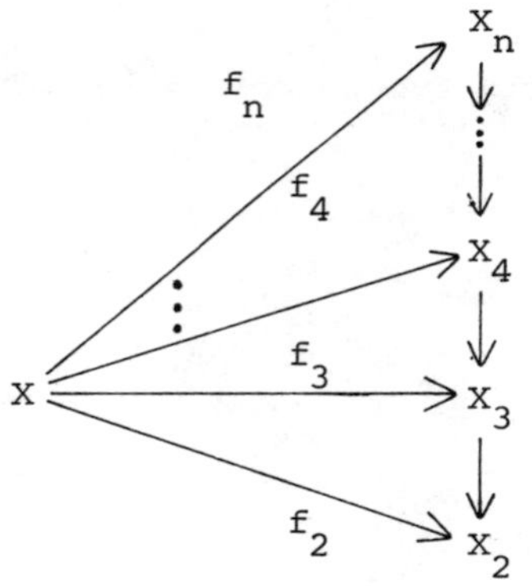

a commutative diagram with:

(1) $\pi_i(X_j) = 0$ for $i > j$,

(2) $X_j \to X_{j-1}$ a principal fibration induced by some map $k^{j+1}\colon X_{j-1} \to K(\pi_j(X),j+1)$, and

(3) $f_j : X \to X_j$ an isomorphism on π_i for all $i \leq j$.

Consider $f_n : X \to X_n$ as an inclusion. The relative homology $H_i(X_n, X)$ is zero for $i \leq n+1$. Furthermore, $H_{n+2}(X_n, X) \cong \pi_{n+2}(X_n, X) \underset{\partial}{\overset{\sim}{\to}} \pi_{n+1}(X)$. By the universal coefficient theorem, $H^{n+2}(X_n, X; \pi_{n+1}(X)) = \mathrm{Hom}(\pi_{n+1}(X), \pi_{n+1}(X))$. Let $\tilde{k}^{n+1} \in H^{n+2}(X_n, X; \pi_{n+1}(X))$ be the class corresponding to the identity homomorphism. This determines a principal fibration and a lifting of f_n

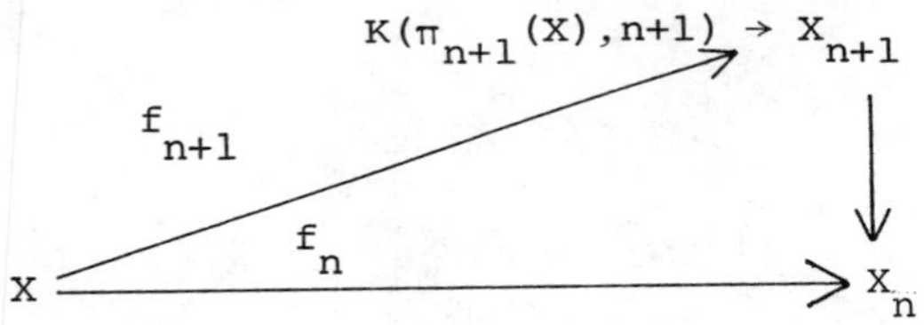

<u>Lemma 7.1</u>: <u>The map</u> $f_{n+1} : X \to X_{n+1}$ <u>meets conditions</u> (1), (2), (3) <u>above</u>.

<u>Proof</u>: Clearly (1) and (2) are satisfied. Also, $(f_{n+1})_* : \pi_i(X) \to \pi_i(X_{n+1})$ is an isomorphism for $i \leq n$. It remains to show that $(f_{n+1})_* : \pi_{n+1}(X) \to \pi_{n+1}(X_{n+1})$ is an isomorphism. We have a map of $(X_n, X) \to (X_n, X_{n+1})$. If this map induces an isomorphism $\pi_{n+2}(X_n, X) \overset{\sim}{\to} \pi_{n+2}(X_n, X_{n+1})$, then it follows that $(f_{n+1})_*$ is an isomorphism on π_{n+1}. Under the identification of $\pi_{n+2}(X_n, X_{n+1})$ with $\pi_{n+1}(K(\pi_{n+1}(X), n+1))$ $= \pi_{n+1}(X)$, the map becomes evaluation of the cohomology class $\tilde{k}^{n+2}$. By definition this map is an isomorphism. ■

Remarks: (1) For each n let X_n' be the CW complex obtained from X by inductively attaching cells of dimension $\geq n+1$ so as to all homotopy groups in dimensions $\geq n+1$. The inclusion $X \subset X_n$ induces an isomorphism on π_i for $i \leq n$. If $f_n: X \to X_n$ is the nth stage of a Postnikov tower for X, then a simple application of obstruction theory shows that f_n extends to a map $f_n': X_n' \to X_n$. This map induces an isomorphism on all homotopy groups.

If $g_n: X \to Y_n$ is the nth stage of another Postnikov tower for X, then we have a commutative diagram

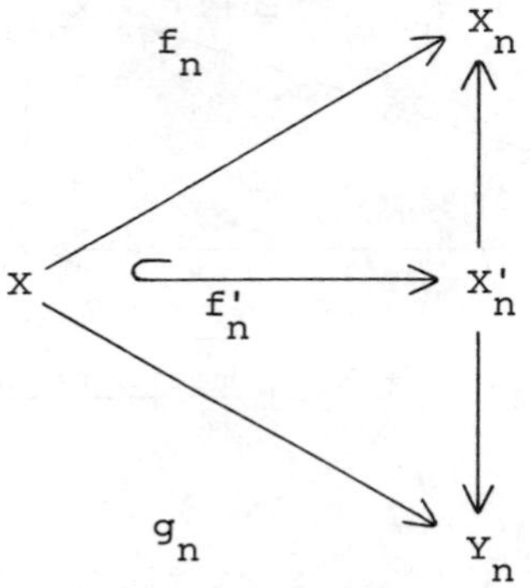

with both vertical arrows being weak homotopy equivalences. It is easy to see that under the resulting identifications of $H^{n+2}(X_n, X; \pi_{n+1}(X))$ with $H^{n+2}(Y_n, X; \pi_{n+1}(X))$ the k-invariants for the $(n+1)^{st}$-stages of the two towers correspond. It is in this sense that the Postnikov tower is unique.

(2) If we form $\lim\{X_i\}$, defined as the subspace of $\pi_{i=2}^{\infty} X_i$ consisting of all compatible sequences, then the maps

$\{f_n: X \to X_n\}$ determine $\lim f_n: X \to \lim\{X_i\}$. This map induces an isomorphism on all homotopy groups. To prove this, one shows that $\pi_j(\lim X_n) = \lim \pi_j(X_n) = \pi_j(X_N)$ for $N \geq j$. It is not true in general for inverse systems that taking homotopy groups commutes with taking inverse limits. It is, however, true for this inverse system since $\pi_{j+1}(X_N) \to \pi_{j+1}(X_{N-1})$ is onto for all N.

If Y is a CW complex and $\varphi: Y \to \lim X_n$ induces an isomorphism on all homotopy groups, then the obstructions to lifting, up to homotopy, $\lim f_n: X \to \lim X_n$ to Y lie in $H^*(X; \pi_*(\lim X_n, Y)) = 0$. Thus there is a map $\psi: X \to Y$ such that $\varphi \circ \psi$ is homotopic to $\lim f_n$. In particular ψ induces an isomorphism on all homotopy groups, and hence ψ is a homotopy equivalence.

This shows how to recover X, up to homotopy equivalence, from a Postnikov tower: it is the unique CW complex, up to homotopy equivalence, which maps to $\lim X_n$ inducing an isomorphism on the homotopy groups.

(3) Suppose X is a CW complex of dimension n. Once we have built the Postnikov tower for X through dimension n, the process of finishing the tower is formal (if complicated). Namely $X \xrightarrow{f_n} X_n$ induces an isomorphism on H_i for $i \leq n$ and is surjective on H_{n+1}. Thus $H_{n+1}(X_n) = 0$, but $H_{n+2}(X_n)$ may not be zero. Suppose it is π. Then the $(n+1)$st homotopy group of X is π and the k-invariant $k^{n+2} \in H^{n+2}(X_n, \pi)$ is the identity homomorphism. By the Serre spectral sequence for

$$\begin{array}{ccc} K(\pi_{n+1}, n+1) & \longrightarrow & X_{n+1} \\ & & \downarrow \\ & & X_n \end{array}$$

one has $H_{n+2}(X_{n+1}) = 0$. We continue in this fashion:

$H_{n+k+1}(X_{n+k}) = 0$, $H_{n+k+2}(X_{n+k}) = \pi_{n+k+1}(X)$, and the k-invariant is the identity map:

$$H_{n+k+2}(X_{n+k}) \longrightarrow \pi_{n+k+1}(X).$$

This ability to complete the Postnikov tower for X once we have X_n without making further reference to the space X is referred to by saying that, after X_n, the process becomes <u>formal</u>. Thus, all simply connected spaces X of homological dimension n are formal after X_n (e.g. there is an algorithm to calculate $\pi_i(X)$, $i > n$, from X_n).

<u>Example</u>: $X = S^2$. In this case $X_2 = K(\mathbb{Z},2) = \mathbb{C}P^\infty$, $H_3(\mathbb{C}P^\infty) = 0$ and $H_4(\mathbb{C}P^\infty) = \mathbb{Z}$. Hence, $\pi_3(S^2) = \mathbb{Z}$. If we form

$$\begin{array}{ccc} K(\mathbb{Z},3) & \longrightarrow & X_3 \\ & & \downarrow \\ & & \mathbb{C}P^\infty \end{array}$$

with k-invariant the identity $H_4(\mathbb{C}P^\infty) \to \mathbb{Z}$, then $H_4(X_3) = 0$ and $H_5(X_3) = \mathbb{Z}/2$. Hence, $\pi_4(S^2) = \mathbb{Z}/2$. Continuing in this way gives a (theoretical) algorithm for calculating all the higher homotopy groups of S^2. This calculation has never been done and seemingly is impossibly complicated to do. It is an amazing theorem of E. Curtis that $\pi_i(S^2) \neq 0$ for all $i \geq 2$.

<u>Rational homotopy theory for simply connected spaces</u>.

We begin with a little of the theory of $\mathbb{Q}$ and $\mathbb{Q}$-vector spaces. Let A be an abelian group (usually infinitely generated). Then A may be given the structure of a $\mathbb{Q}$ vector space if and only if $A \simeq A \otimes_{\mathbb{Z}} \mathbb{Z} \xrightarrow{\sim} A \otimes_{\mathbb{Z}} \mathbb{Q}$. (This is equivalent to the equation

$\alpha x = \beta$ having a unique solution for $\alpha \in \mathbb{Z} - \{0\}$ and $\beta \in A$.)

If $0 \to A_1 \to A_2 \to A_3 \to 0$ is a short exact sequence then so is $0 \to A_1 \otimes_{\mathbb{Z}} \mathbb{Q} \to A_2 \otimes_{\mathbb{Z}} \mathbb{Q} \to A_3 \otimes \mathbb{Q}_{\mathbb{Z}} \to 0$.

<u>Lemma 7.2</u>: (a) *If* $0 \to A_1 \to A_2 \to A_3 \to 0$ *is a short exact sequence, then if* 2 *of the* 3 *terms are* $\mathbb{Q}$-*vector spaces so is the* 3rd.

(b) *If* A *is an abelian group and has a composition series* $A = A_0 \supset A_1 \supset \ldots \supset A_n \supset 0$ *with successive quotients* $\mathbb{Q}$-*vector spaces, then* A *is a* $\mathbb{Q}$-*vector space*.

(c) $H_*(X,\mathbb{Q}) \cong H_*(X) \otimes_{\mathbb{Z}} \mathbb{Q}$; $H^*(X,\mathbb{Q}) \cong \mathrm{Hom}_{\mathbb{Z}}(H_*(X),\mathbb{Q}) \cong [H_*(X,\mathbb{Q})]^*$.

(d) *If* $\tilde{H}_*(X)$ *is a* $\mathbb{Q}$-*vector space, then* $\tilde{H}_*(X;G)$ *is a* $\mathbb{Q}$-*vector space for any abelian group* G.(†)

<u>Proof:</u> (a):

$$\begin{array}{ccccccccc} 0 & \longrightarrow & A_1 & \longrightarrow & A_2 & \longrightarrow & A_3 & \longrightarrow & 0 \\ & & \downarrow & & \downarrow & & \downarrow & & \\ 0 & \longrightarrow & A_1 \otimes \mathbb{Q} & \longrightarrow & A_2 \otimes \mathbb{Q} & \longrightarrow & A_3 \otimes \mathbb{Q} & \longrightarrow & 0 \end{array}$$

is a commutative ladder of short exact sequences. By hypothesis 2 of the 3 vertical maps are isomorphism's. Thus by the 5 lemma (cf. J. Clayburn, It's my turn, 1980) so is the 3rd.

(b): Follows immediately by induction from (a).

(c) and (d) are left to the reader as exercises. ∎

<u>Corollary 7.3</u>: $H_i(X;\mathbb{Q})$ *and* $H^i(X,\mathbb{Q})$ *are* $\mathbb{Q}$-*vector spaces*.

(†) $\tilde{H}_*$ = reduced homology groups.

<u>Definition</u>: A $\mathbb{Q}$-<u>space</u> is a space, X, satisfying:

(1) X is homotopy equivalent to a CW complex (generally having ∞-many cells in each dimension).

(2) $\pi_1(X) = 0$.

(3) $\pi_*(X)$ is a $\mathbb{Q}$-vector space for all $* \geq 1$.

<u>Alternate definition</u>: A space X satisfying (1) and (2) above is a $\mathbb{Q}$-space if in addition (3) $\widetilde{H}_*(X,\mathbb{Z})$ is a $\mathbb{Q}$-vector space.

<u>Theorem 7.4</u>: <u>The two definitions of a $\mathbb{Q}$-space are equivalent</u>.

<u>Proof</u>: We will need the following lemma.

<u>Lemma 7.5</u>: (a) $\widetilde{H}_*(K(\mathbb{Q},n),\mathbb{Z})$ <u>is a $\mathbb{Q}$-vector space</u>.

(b) $H^*(K(\mathbb{Q},2n);\mathbb{Q})$ <u>is a $\mathbb{Q}$-polynomial algebra on one generator of degree</u> 2n.

(c) $H^*(K(\mathbb{Q},2n+1);\mathbb{Q})$ <u>is a $\mathbb{Q}$-exterior algebra on one generator of degree</u> 2n + 1.

<u>Proof:</u> We construct $K(\mathbb{Q},1)$ as an iterated mapping cylinder:

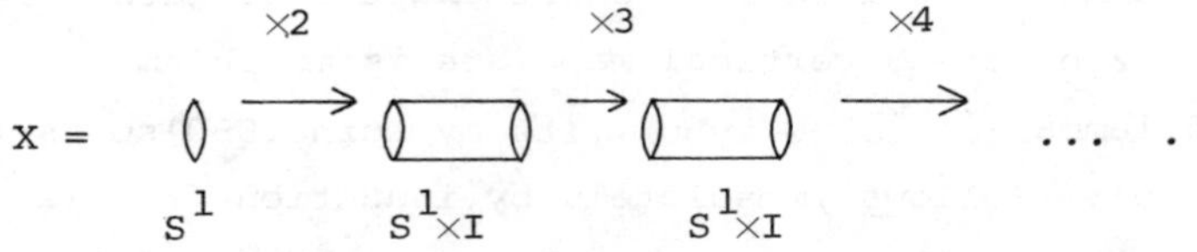

Since homotopy commutes with direct limits,

$$\pi_1(X) = \lim_{\rightarrow}\{\mathbb{Z}, \times n\} = \mathbb{Q},$$

and

$$\pi_i(X) = \lim_{\rightarrow}\{0\} = 0 \quad \text{for} \quad i > 1.$$

Thus, X is a $K(\mathbb{Q},1)$. Since homology also commutes with direct limits

$$\tilde{H}_*(X;\mathbb{Z}) = \begin{cases} \mathbb{Q} & \text{for } * = 1 \\ 0 & \text{for } * \neq 1. \end{cases}$$

Suppose inductively that we have shown that $\tilde{H}_*(K(\mathbb{Q},n-1);\mathbb{Z})$ is a $\mathbb{Q}$-vector space. Consider the Serre spectral sequence for

$$\begin{array}{ccc} K(\mathbb{Q},n-1) & \longrightarrow & \mathcal{P} \\ & & \downarrow \\ & & K(\mathbb{Q},n) \end{array}$$

with $\mathbb{Z}$-coefficients and with $\mathbb{Q}$-coefficients.

There is a comparison theorem for spectral sequences which says that, given two spectral sequences and a map between them inducing an isomorphism on $E_\infty^{*,*}$ for $(*,*) \neq (0,0)$ and on $E_2^{0,q}$ for all $q > 0$, then it is an isomorphism on $E_2^{p,q}$ for all $(p,q) \neq (0,0)$. In particular, it is an isomorphism on $E_2^{p,0}$ for $p>0$. This result is an algebraic one proved by induction, cf. exercise 42.

Applying it to the Serre spectral sequences for the above fibration with $\mathbb{Z}$- and $\mathbb{Q}$-coefficients we see that if $\tilde{H}_*(K(\mathbb{Q},n-1);\mathbb{Z})$ is a $\mathbb{Q}$-vector space, then so is $\tilde{H}_*(K(\mathbb{Q},n);\mathbb{Z})$.

Parts (b) and (c) are proved by the same inductive argument given in section IV to calculate $H^*(K(\mathbb{Z},n);\mathbb{Q})$. ∎

Corollary 7.6: $K(\mathbb{Z},n) \to K(\mathbb{Q},n)$ induces an isomorphism on rational cohomology, and thus on rational homology.

Using lemma 7.5 we will show that, if $\pi_1(X) = 0$ and $\pi_1(X)$ is a $\mathbb{Q}$-vector space for all i, then $\tilde{H}_i(X)$ is a $\mathbb{Q}$-vector space for all i. Since $H_*(X) \cong H_*(X_n)$ ($X_n = n^{th}$ stage in the Postnikov system for X) for $* \leq n$, it suffices to prove by induction that $\tilde{H}_*(X_n)$ is a $\mathbb{Q}$-vector space. Since $X_2 = K(\pi_2(X),2)$ and $\pi_2(X) \cong \oplus\mathbb{Q}$, from lemma 7.5 we see that $\tilde{H}_*(X_2)$ is a $\mathbb{Q}$-vector space. Suppose $\tilde{H}_*(X_{n-1})$ is a $\mathbb{Q}$-vector space and consider the Serre spectral sequence for

$$\begin{array}{ccc} K(\pi_n(X),n) & \longrightarrow & X_n \\ & & \downarrow \\ & & X_{n-1} \end{array}$$

Now $E_2^{p,q}$ is a rational vector space for all $(p,q) \neq (0,0)$, and consequently E_∞ is also. Thus there are composition series for $\tilde{H}_i(X_n)$ with successive quotients $\mathbb{Q}$-vector spaces. This proves $\tilde{H}_i(X_n)$ is a $\mathbb{Q}$-vector space for all i.

Conversely, suppose we have a space X with $\pi_1(X) = 0$ and $\tilde{H}_*(X)$ a $\mathbb{Q}$-vector space. We will show by induction that $\pi_i(X)$ is a $\mathbb{Q}$-vector space. Suppose we know that $\pi_*(X_{n-1})$ are $\mathbb{Q}$-vector spaces. Consider the (homology) Serre spectral sequence for

$$\begin{array}{ccc} K(\pi_n(X),n) & \longrightarrow & X_n \\ & & \downarrow \\ & & X_{n-1}\ . \end{array}$$

The E^2-term looks like

$\pi_n(X)$	*	*	*...*
0			
⋮	0		
0			
$\mathbb{Z}$	$\mathbb{Q}$-vector spaces		

Let $B_n \subset \pi_n(X)$ be the image of $d_{n+1}: E^{n+1}_{n+1,0} \to E^{n+1}_{0,n} = \pi_n(X)$. Since $E^{n+1}_{n+1,0}$ is a $\mathbb{Q}$-vector space, so is B_n. The composition series for $H_n(X_n)$ is

$$0 \longrightarrow \pi_n(X)/B_n \longrightarrow H_n(X_n) \longrightarrow E^2_{n,0} \longrightarrow 0.$$

Since $H_n(X_n) = H_n(X)$, it is a $\mathbb{Q}$-vector space. Hence, $\pi_n(X)/B_n$ is a $\mathbb{Q}$-vector space. Consequently, so is $\pi_n(X)$. ■

Theorem 7.7: Let X and $X_{(0)}$ be simply connected CW complexes with $X_{(0)}$ a $\mathbb{Q}$-space and $f: X \to X_{(0)}$. The following three conditions are equivalent:

(a) $f_*: \tilde{H}_*(X,\mathbb{Q}) \to \tilde{H}_*(X_{(0)};\mathbb{Q}) = \tilde{H}_*(X_{(0)})$ is an isomorphism

(b) $f_\#: \pi_*(X) \otimes \mathbb{Q} \to \pi_*(X_{(0)}) \otimes \mathbb{Q} = \pi_*(X_{(0)})$ is an isomorphism

(c) f is universal for maps of X into $\mathbb{Q}$-space; i.e., given $g: X \to Y_{(0)}$ with $Y_{(0)}$ a $\mathbb{Q}$-space, then g factors uniquely up to homotopy through $X_{(0)}$

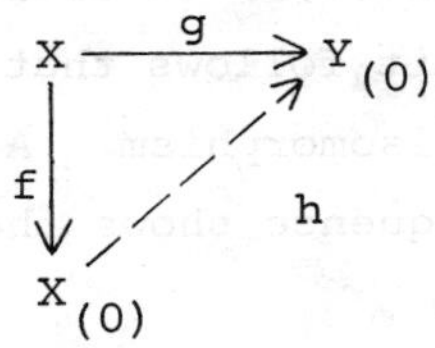

Proof: (a) ⇒ (c): Given f satisfying (a) and $g: X \to Y_0$, then the obstructions to extending g over all of $X_{(0)}$ (considering f now as an inclusion) lie in $H^*(X_{(0)},X;\pi_*(Y_{(0)}))$. Since $f_*: H_*(X) \to H_*(X_{(0)})$ is an isomorphism with $\mathbb{Q}$-coefficients and $\pi_*(Y_{(0)})$ is a $\mathbb{Q}$-vector space $H^{*+1}(X_{(0)},X;\pi_*(Y_0)) = 0$. Thus such an h exists. The obstruction to any two h's being homotopic relative to g

are in $H^*(X_{(0)},X;\pi_*(Y_{(0)})) = 0$: this proves a) ⇒ c).

c) ⇒ a) Apply universality to $Y_{(0)} = K(\mathbb{Q},n)$ and we see that there is a 1-1 correspondence

$$\begin{array}{ccc} [X,K(\mathbb{Q},n)] & \overset{f^*}{\longleftrightarrow} & [X_{(0)},K(\mathbb{Q},n)] \\ \updownarrow & & \updownarrow \\ H^n(X,\mathbb{Q}) & \overset{f^*}{\longleftarrow} & H^n(X_{(0)};\mathbb{Q}) \end{array}$$

and thus f^* is an isomorphism on $\mathbb{Q}$-cohomology and thus on $\mathbb{Q}$-homology.

To prove a) ⇔ b) we need the following lemma.

<u>Lemma 7.8</u>: <u>If</u> X <u>is a simply connected CW complex, then</u> $\pi_*(X) \otimes \mathbb{Q} = 0$ <u>if and only if</u> $\widetilde{H}_*(X;\mathbb{Q}) = 0$.

(This is a variant of the Hurewicz theorem modulo the Serre class of abelian groups A such that $A \otimes \mathbb{Q} = 0$.)

<u>Proof Step 1</u>: <u>If</u> $\pi = \mathbb{Z}/k\mathbb{Z}$, <u>then</u> $\widetilde{H}_*(K(\pi,n);\mathbb{Q}) = 0$.

We have a fibration $K(\pi,n) \to K(\mathbb{Z},n+1) \xrightarrow{\cdot k} K(\mathbb{Z},n+1)$. From the fact that $H^*(K(\mathbb{Z},n+1);\mathbb{Q})$ is either a polynomial algebra or an exterior algebra, it follows that $(\cdot k): H^*(K(\mathbb{Z},n+1);\mathbb{Q}) \to H^*(K(\mathbb{Z},n+1);\mathbb{Q})$ is an isomorphism. A simple application of the Serre spectral sequence shows that this implies $\widetilde{H}^*(K(\pi,n);\mathbb{Q}) = 0$.

<u>Step 2</u>: $\pi \otimes \mathbb{Q} = 0 \Rightarrow \widetilde{H}_*(K(\pi,n),\mathbb{Q}) = 0$.

If π is a sum of cyclic groups, then the result follows from Step 1. In general π is a limit of sums of cyclic groups. Since homology commutes with limits, step 2 is true for all groups π.

<u>Step 3</u>: <u>If</u> $\pi_i(X) \otimes \mathbb{Q} = 0$ <u>for all</u> i <u>and</u> $\pi\,(X) = 0$, <u>then</u>

$\widetilde{H}_i(X;\mathbb{Q}) = 0$.

We build the Postnikov tower for X and prove the result inductively for the X_n. Suppose we know that $\widetilde{H}_*(X_{n-1};\mathbb{Q}) = 0$. Consider the Serre spectral sequence for $K(\pi_n(X),n) \to X_n \to X_{n-1}$ with $\mathbb{Q}$-coefficients. $E^2_{p,q} = 0$ for all $(p,q) \neq (0,0)$. Hence $E^\infty_{p,q} = 0$ for all $(p,q) \neq (0,0)$. Thus, $\widetilde{H}_i(X_n;\mathbb{Q}) = 0$ for all i. Since $H_i(X_n) = H_i(X)$ if $n > i$, this establishes step 3.

<u>Step 4:</u> <u>If</u> $\pi_1(X) = 0$ <u>and</u> $\widetilde{H}_i(X;\mathbb{Q}) = 0$ <u>for all</u> i, <u>then</u> $\pi_i(X) \otimes \mathbb{Q} = 0$ <u>for all</u> i.

Consider the Serre spectral sequence for

$$\begin{array}{ccc} K(\pi_n(X),n) & \longrightarrow & X_n \\ & & \downarrow \\ & & X_{n-1} \end{array}$$

with $\mathbb{Q}$-coefficients. At the E^2-term we have

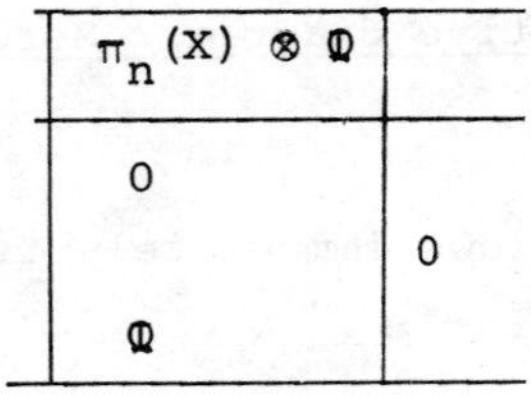

Thus, $\pi_n(X) \otimes \mathbb{Q} = E^\infty_{0,n} = H_n(X;\mathbb{Q})$. But $H_n(X_n) \cong H_n(X)$ and by assumption $H_n(X;\mathbb{Q}) = 0$. Thus $\pi_n(X) \otimes \mathbb{Q} = 0$. This completes the lemma. ■

From this we prove (a) ⇔ (b). Namely let $f: X \to Y$ and let F_f be the homotopy theoretic fiber. Then $f_*: \pi_*(X) \otimes \mathbb{Q} \to \pi_*(Y) \otimes \mathbb{Q}$ is an isomorphism ⇔ $\pi_*(F_f) \otimes \mathbb{Q} = 0$ ⇔ $H_*(F_f;\mathbb{Q}) = 0$ ⇔ $H_*(X;\mathbb{Q}) \xrightarrow{f_*} H_*(Y;\mathbb{Q})$ is an isomorphism. (This last equivalence comes from the Serre spectral sequence.) This completes the proof of the theorem. ■

<u>Definition</u>: Given X and $f\colon X \to X_{(0)}$ with $X_{(0)}$ a $\mathbb{Q}$-space and f satisfying (1), (2), or (3) above (and hence all of them), we call $f\colon X \to X_{(0)}$ the <u>localization at</u> 0 <u>of</u> X.

The terminology comes from the fact that if we localize $\mathbb{Z}$ at 0 we get $\mathbb{Q}$.

We will not, in this course, consider any other localization; although it is possible to localize at any prime ideal. In fact, there is a Hasse-Minkowski principle which allows one to recover the whole space from its various localizations.

<u>Theorem 7.9</u>: <u>If</u> $\varphi\colon X \to X_{(0)}$ <u>and</u> $\varphi'\colon X \to X'_{(0)}$ <u>are localizations of</u> X, <u>then there is a homotopy equivalence</u> $h\colon X_{(0)} \to X'_{(0)}$ <u>such that</u>

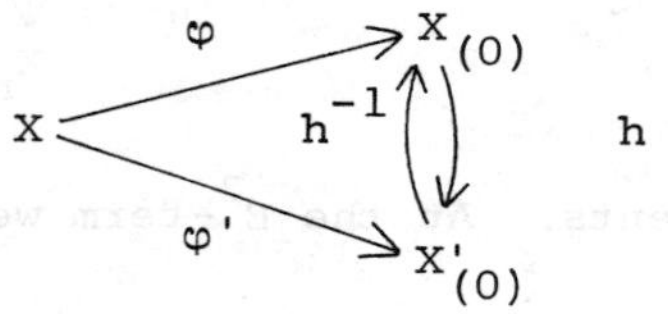

<u>is a homotopy commutative diagram. Moreover</u>, h <u>is unique up to homotopy</u>.

<u>Proof</u>: The proof follows immediately from the universal property of localization. ■

<u>Construction of the localization of a space</u>.

The construction of the localization of a space goes by induction on the Postnikov tower of the space. We will assume that X is a CW complex and is simply connected. The idea of the proof is to tensor both the groups and the k-invariant with $\mathbb{Q}$.

Suppose inductively that we have a localization

$X_{n-1} \xrightarrow{\varphi_{n-1}} X_{n-1}(0)$. Then:

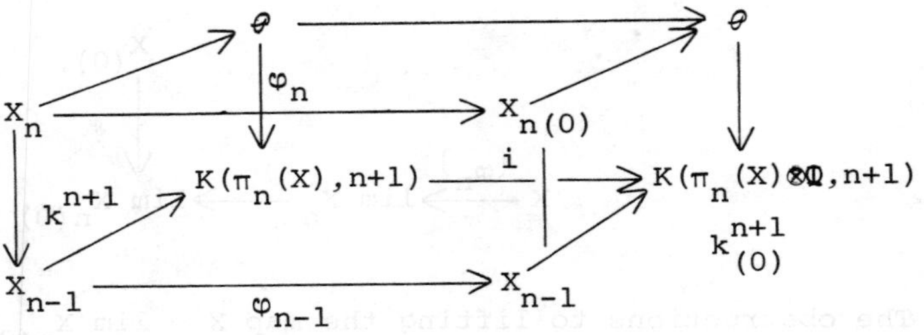

Since $K(\pi_n(X) \otimes \mathbb{Q}, n+1)$ is a $\mathbb{Q}$-space, $i \circ k^{n+1}$ factors uniquely through $X_{n-1(0)}$. Let $X_{n(0)}$ be the fibration induced over $X_{n-1(0)}$ from $k^{n+1}_{(0)}$. Since the total compositions of

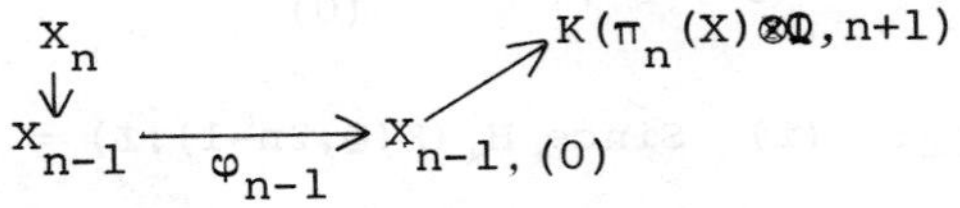

and

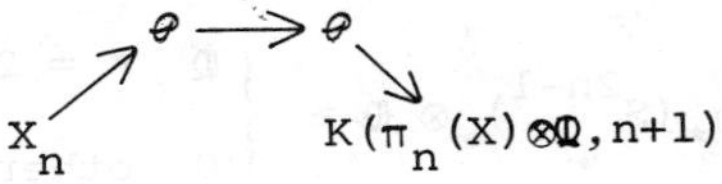

are the same map, this defines $\varphi_n\colon X_n \to X_{n(0)}$. By the commutativity of the diagram we see that $(\varphi_n)_*\colon \pi_n(X) \to \pi_n(X_{n(0)})$ is an isomorphism when tensored with $\mathbb{Q}$. This proves that $\varphi_n\colon X_n \to X_{n(0)}$ is the localization of X_n at 0. To complete the proof we need the following lemma.

<u>Lemma 7.10</u>: <u>Let</u> Y <u>be a topological space. There is a CW complex</u> X <u>and a map</u> $\psi\colon S \to Y$ <u>which induces an isomorphism on all homotopy groups. If</u> $\psi'\colon X' \to Y$ <u>is another such, then there is a homotopy equivalence</u> $h\colon X \to X'$ <u>so that</u> $\psi' \circ h$ <u>is homotopic to</u> ψ.

The proof of this lemma is left as an exercise for the reader.

We apply the lemma with $Y = \varprojlim X_n(0)$. Let $X_{(0)}$ be a CW complex with $\psi\colon X_{(0)} \to Y$. The maps $X_n \to X_{n(0)}$ define a map $\varprojlim X_n \to \varprojlim X_{n(0)}$. Thus, we have

$$\begin{array}{ccccc} & & & & X_{(0)} \\ & & & & \downarrow \psi \\ X & \xrightarrow{\{\varphi_n\}} & \lim X_n & \longrightarrow & \lim X_{n(0)} \end{array}$$

The obstructions to lifting the map $X \to \varprojlim X_{n(0)}$ to a map $X \to X_{(0)}$ lie in $H^*(X;\pi_{*-1}(F))$ where F is the homotopy-theoretic fiber of ψ. Since ψ induces an isomorphism on homotopy groups, $\pi_*(F) = 0$.

The resulting map $\varphi\colon X \to X_{(0)}$ is a localization at 0.

<u>Calculations</u>: (i) Since $H_*(K(\mathbb{Q},2n-1);\mathbb{Z}) = \{\begin{smallmatrix}\mathbb{Q} & * = 2n-1 \\ 0 & \text{otherwise}\end{smallmatrix}$, $S^{2n-1} \to K(\mathbb{Q},2n-1)$ is a localization. Thus

$$\pi_*(S^{2n-1}) \otimes \mathbb{Q} \cong \begin{cases} \mathbb{Q} & * = 2n-1 \\ 0 & \text{otherwise} \end{cases}.$$

Since $\pi_i(S^{2n-1})$ is always finitely generated

$$\pi_i(S^{2n-1}) \cong \begin{cases} \mathbb{Z} & i = 2n-1 \\ \text{finite group} & i \neq 2n-1 \end{cases}$$

(ii) $S^{2n} \to K(\mathbb{Q},2n)$ induces an isomorphism in rational cohomology through degree $(4n-1)$. The kernel of $H^{4n}(K(\mathbb{Q},2n);\mathbb{Q}) \to H^{4n}(S^{2n};\mathbb{Q})$ is generated by $(\iota)^2$. Form the principle fibration $K(\mathbb{Q},4n-1) \to E \to K(\mathbb{Q},2n)$ with k-invariant ι^2. A calculation using the Serre spectral sequence shows that $H^*(E;\mathbb{Q}) \cong H^*(S^{2n};\mathbb{Q})$. Furthermore, we can lift $S^{2n} \to K(\mathbb{Q},2n)$ to a map $S^{2n} \to E$. This proves that

$$\pi_i(S^{2n}) = \begin{cases} \mathbb{Z} & i = 2n \\ \mathbb{Z} \oplus \text{finite group} & i = 4n-1 \\ \text{finite group} & i \neq 2n \text{ or } 4n-1 \end{cases}$$

(iii) There is a map $\mathbb{C}P^n \to K(\mathbb{Q},2)$ which induces an isomorphism on rational cohomology through degree $2n + 1$. Let $E \to K(\mathbb{Q},2)$ be a principal fibration with fiber $K(\mathbb{Q},2n-1)$ and k-invariant ι^{n+1}. As before one sees that $\mathbb{C}P^n \to E$ induces an isomorphism on rational cohomology. Thus, E is $\mathbb{C}P^n_{(0)}$.

(iv) We have seen that $H^*(BU(n);\mathbb{Q})$ is a polynomial algebra generated by $c_1,\ldots,c_n$. Define

$$BU(n) \xrightarrow[(c_1,\ldots,c_n)]{} \Pi_{k=1}^{n} K(\mathbb{Q},2k).$$

This map induces an isomorphism on rational cohomology. Thus $BU(n)_{(0)} \cong \Pi_{k=1}^{n} K(\mathbb{Q},2k)$. This means that all the k-invariants of $BU(n)$ are trivial when tensored with $\mathbb{Q}$. This implies that they are all of finite order. This also shows the rational Bott periodicity theorem: $\Omega^2 BU_{(0)} \cong BU_{(0)}$.

These examples illustrate a recurring theme--homotopy theory over $\mathbb{Q}$ is much simpler than homotopy theory over $\mathbb{Z}$. We have a chance of getting complete answers over $\mathbb{Q}$ whereas over $\mathbb{Z}$ this is seldom possible; e.g., $\pi_*(S^n)$.

VIII. deRham's theorem for simplicial complexes

In the next 4 sections we will see how to calculate the rational homotopy type of a space from certain differential forms.

A) Piecewise linear forms (P.L. forms).

We will work on a simplicial complex K . Recall that K is the union of n-simplices Δ^n, where Δ^n may be thought of as

$$\Delta^n = \{(t_0,\ldots,t_n): 0 \leq t_i \leq 1, \Sigma_{i=0}^n t_i = 1\}.$$

(The t_i are called barycentric coordinates.) The first few pictures are

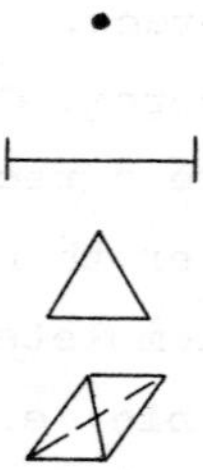

0-simplex

1-simplex

2-simplex

3-simplex .

Note that $\partial\Delta^n$ is a union of (n-1)-simplices, and is topologically an S^{n-1}.

Consider the restriction to Δ^n of all forms in $\mathbb{R}^{n+1}$ of the form

$$\Sigma\, \varphi_{i_1\ldots i_j}\, dt_{i_j} \wedge\ldots\wedge dt_{i_j}$$

where the $\varphi_{i_1\ldots i_j}$ are polynomials with $\mathbb{Q}$-coefficients.

There is the relation $\Sigma_{i=0}^{n} t_i = 1$ and the derived relation is $dt_0 + \ldots + dt_n = 0$. We call this algebra $A^*(\Delta^n)$. If $\Delta^k \subset \Delta^n$ is a face, then there is a restriction map $A^*(\Delta^n) \to A^*(\Delta^k)$.

Let K be a simplicial complex. Define

$$A^*(K) = \{(\omega_\sigma)_{\sigma \in K} : \omega_\sigma \in A^*(\sigma) \text{ and } \omega_\sigma | \tau = \omega_\tau \text{ if } \tau \text{ is a face of } \sigma\}.$$

Thus, $A^*(K)$ is the collection of forms, one on each simplex of K, which are compatible under restriction to faces. Clearly, wedge product and d, both defined by the corresponding operations on each simplex, give $A^*(K)$ the structure of a differential graded algebra (D.G.A.). It is defined over $\mathbb{Q}$.

There is a map:

$$A^*(K) \xrightarrow[\rho]{} C^*(K;\mathbb{Q})$$

defined by $\langle \rho(\omega), \Delta^n \rangle = \int_{\Delta^n} \omega$. This map is a map of cochain complexes by Stokes' theorem (which is valid in our setting).

<u>Theorem (P.L. deRham Theorem):</u> ρ <u>induces an algebra isomorphism on cohomology</u>.

We will deduce the P.L.D.R. theorem from the following proposition.

<u>Proposition 8.1</u>: (i) <u>Let</u> $\varphi \in A^n(K$ <u>satisfy</u> $d\varphi = 0$, $\rho(\varphi) = 0$ (<u>i.e.</u> $\int_{\Delta^n} \varphi = 0$ <u>for all</u> Δ^n). <u>Then there exists</u> $\psi \in A^{n-1}(K)$ <u>such that</u> $d\psi = \varphi$, $\rho(\psi) = 0$.

(ii) $A^*((K) \xrightarrow{\rho} C^*(K)$ <u>is onto</u>.

<u>Proof of P.L.D.R. assuming proposition 8.1 (additive statement):</u> We have by (ii)

$$0 \longrightarrow B^*(K) \longrightarrow A^*(K) \xrightarrow{\rho} C^*(K) \longrightarrow 0.$$

The first part of the proposition says that

$$H^*(B^*(K)) = 0,$$

and thus

$$H^*_{DR}(K) \cong H^*(K;\mathbb{Q}).$$

The multiplicative statement will be considered later.

B) Lemmas about P.L. forms.

Poincaré Lemma 8.2: Let $c(K)$ be the simplicial complex which is the cone over a finite complex K. Suppose φ^{ℓ} is a closed form in $A^*(c(K))$, $\ell > 0$. Then $\varphi^{\ell} = d\psi^{\ell-1}$ for some $\psi^{\ell-1} \in A^*(c(K))$.

Proof: $c(K)$ is the join of the point c with K. Points in $c(K)$ are denoted by $sk + (1-s)\cdot c$ where $0 \leq s \leq 1$ and $k \in K$, with the proviso that $0\cdot k+1\cdot c = c$ for all $k \in K$. Define

$$\mu\colon c(K) \times I \to c(K)$$

by

$$\mu(s\cdot k+(1-s)\cdot c,t) = s(1-t)\cdot k + (s+t-st)\cdot c.$$

If $\lambda \in A^*(c(K))$, then $\rho^*(\lambda)$ is a form on $c(K) \times I$. Restricted to any $c(\sigma) \times I$, σ a simplex of K, $\rho^*(\lambda)$ is a polynomial form with $\mathbb{Q}$ coefficients when expressed in terms of the barycentric coordinates of σ, s, and t. These forms patch together.

<u>Claim:</u> If φ^ℓ is a closed form of degree $\ell > 0$ on $c(K)$, then $d(-\int_{t=0}^{t=1} \rho^*(\varphi^\ell)) = \varphi^\ell$.

<u>Explanation</u>: If we expand $\rho^*(\varphi^\ell)|c(\sigma) \times I$ as $\Sigma_{i=1}^N \alpha_i(\sigma)t^i + \beta_i(\sigma)t^i dt$ where $\alpha_i(\sigma)$ and $\beta_i(\sigma)$ are in $A^*(c(\sigma))$, then the $\alpha_i(\sigma)$ and $\beta_i(\sigma)$ patch together to give forms α_i and β_i in $A^*(c(K))$. If

$$\rho^*\varphi^\ell = \Sigma_{i=0}^N \alpha_i t^i + \beta_i t^i dt \text{ with } \alpha_i, \beta_i \in A^*(c(K)),$$

then we define

$$\int_{t=0}^{t=1} \rho^*\varphi^\ell = \Sigma_{i=0}^N (-1)^{\deg \beta_i} \frac{\beta_i}{i+1} .$$

We now check that

$$d(-\int_{t=0}^{t=1} \rho^*\varphi) = \varphi.$$

Since $\rho|_{t=0}$ is the identity

$$\varphi = \rho^*\varphi|_{\substack{t=0\\ dt=0}} = \alpha_0.$$

Since $\rho|_{t=1}$ is the constant map to c

$$\rho^*\varphi|_{\substack{t=1\\ dt=0}} = \Sigma_{i=0}^N \alpha_i = 0.$$

(Here is the only place we use $\ell > 0$.)

Lastly, since φ is closed, $\rho^*(d\varphi) = d\rho^*\varphi = 0$; i.e.,

$$\begin{cases} d\,\alpha_i = 0 & \text{for } i \geq 0, \text{ and} \\ (-1)^{\deg \alpha_i} i\alpha_i + d\beta_{i-1} = 0 & \text{for } i \geq 1 \end{cases}$$

Hence,
$$d\left(-\int_{t=0}^{t=1} \rho^*\varphi\right) = \Sigma_{i=0}^{N} \left(\frac{(-1)^{\deg \beta_i+1}}{i+1}\right) d\beta_i$$

$$= - \Sigma_{i=1}^{N} \alpha_i = \alpha_0 = \varphi. \blacksquare$$

<u>Extension Lemma 8.3</u>: <u>Let</u> φ^ℓ <u>be a form in</u> $A^*(\partial\Delta^n)$. <u>There is a form</u> $\tilde{\varphi}^\ell \in A^*(\Delta^n)$ <u>such that</u> $\tilde{\varphi}^\ell|\partial\Delta^n = \varphi$.

<u>Proof</u>: Let σ be an (n-1)-dimensional face of Δ^n, say $\sigma = \{(t_0,\ldots,t_n)|t_n = 0\}$. Let $\alpha \in A^*(\sigma)$. Let v be the vertex $\{t_n = 1\}$ and $U \subset \Delta^n$ be the complement of this vertex. There is a stereographic projection from the vertex $\pi: U \to \sigma$

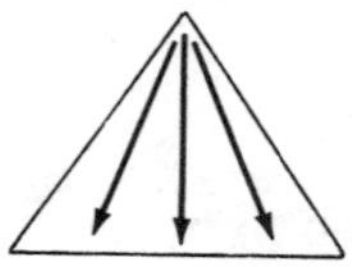

σ

$$\pi(t_0,\ldots,t_n) = \left(\frac{t_0}{1-t_n},\frac{t_1}{1-t_n},\ldots,\frac{t_{n-1}}{1-t_n}\right).$$

The form $\pi^*(\alpha)$, considered as a form on U, is a polynomial form with $\mathbb{Q}$-coefficients in the variables $t_0,\ldots,t_{n-1},1/1-t_n,dt_0,\ldots,dt_{n-1},d(1/1-t_n)$. Since $d(1/1-t_n) = 1/(1-t_n)^2 dt_n$, $\pi^*(\alpha)$ is a polynomial form with $\mathbb{Q}$-coefficients in the variables. $t_0,\ldots,t_{n-1},1/1-t_n$, $dt_0,\ldots,dt_n$. Hence for some $N \geq 0$ $(1-t_n)^N\pi^*\alpha = \tilde{\alpha}$ is a $\mathbb{Q}$-polynomial form on Δ^n. It is the required extension of α to all of Δ^n. Note that if τ is a face of σ and if $\alpha|\tau = 0$, then $\tilde{\alpha}$ restricted to the join of τ and ν is 0.

Now suppose that $\varphi \in A^*(\partial\Delta^n)$. Let σ_0 be the face $\{t_0 = 0\}$, and let $\varphi_0 \in A^*(\sigma_0)$ be the restriction of φ to σ_0. Extend this to a form $\tilde{\varphi}_0$ in Δ^n. The difference $\varphi - \tilde{\varphi}_0|\partial\Delta^n$ vanishes on σ_0. Call this difference φ_1. Let σ_1 be the face $\{t_1 = 0\}$. Extend $\varphi_1|\sigma_1$ to a form $\tilde{\varphi}_1$ on Δ^n. We know that $\tilde{\varphi}_1|\sigma_0 = 0$. Thus, $\varphi - (\tilde{\varphi}_0 + \tilde{\varphi}_1|\partial\Delta^n)$ vanishes on $\sigma_0 \cup \sigma_1$. Continue in this manner defining φ_i on σ_i and $\tilde{\varphi}_i$ on Δ^n. In the end we have $\varphi = \Sigma_{i=0}^n \tilde{\varphi}_i|\partial\Delta^n$.■

<u>Lemma 8.4</u>: (a_n) <u>Let φ^ℓ be a closed form in $A^*(\Delta^n)$ which vanishes on $\partial\Delta^n$. If $\ell = n$, then assume also that</u> $\int_{\Delta^n} \varphi = 0$. <u>Then $\varphi^\ell = d\psi^{\ell-1}$ for some $\psi^{\ell-1}$ which vanishes on $\partial\Delta^n$.</u>

(b_n) <u>Let φ^ℓ be a closed form in $A^*(\partial\Delta^n)$</u>, $\ell > 0$. <u>If</u> $\ell = n-1$, <u>then assume that</u> $\int_{\partial\Delta^n} \varphi = 0$. <u>Then</u> $\varphi = d\psi$ <u>for some</u> $\psi \in A^*(\partial\Delta^n)$.

<u>Proof</u>: Clearly, (a_0) is true,and (b_0) and (b_1) are vacuous.

(a_1) <u>is true</u>:

Since k forms on a n-simplex vanish for $k > n$,to prove (a_1) we need only consider functions and 1-forms. (a_1) is clearly true for functions. For 1-forms it is just the fundamental theorem of calculus. If $p(t)dt$ is a closed 1-form on $[0,1]$, then there is a polynomial $Q(t)$ so that $Q'(t) = p(t)$ and $Q(0) = 0$. Its value at 1 is $\int_0^1 p(t)dt = 0$.

<u>$(a_{n-1}) \Rightarrow (b_n)$</u>

Let φ be a closed form in $A^*(\partial\Delta^n)$. Let $\sigma_n=\{t_n=0\}$be a co-dimension 1 face of Δ^n. By the Poincaré Lemma $\varphi|(\partial\Delta^n\text{-int }\sigma_n) = d\psi$ for some $\psi \in A^*(\partial\Delta^n - \text{int }\sigma_n)$. By the extension lemma we can extend ψ to $\tilde{\psi}$ in $A^*(\partial\Delta^n)$. Then $\varphi - d\tilde{\psi}$ is a closed form on $\partial\Delta^n$ vanishing except on the face $\{t_n = 0\}$. On this face it is a relative form; i.e., it vanishes when restricted to $\partial\sigma^n$. If $\ell = n-1$ then $0 = \int_{\partial\Delta^n} \varphi^n = \int_{\partial\Delta^n} \varphi^n - d\tilde{\psi}$ $= \int_{\{t_n=0\}} \varphi - d\tilde{\psi}$. Thus by (a_{n-1}), $\varphi - d\tilde{\psi} = d\mu$ for some relative form μ on $\{t_n = 0\}$. Let $\tilde{\mu}$ be the extension by 0 of μ to the rest of $\partial\Delta^n$. Then $\varphi = d(\tilde{\psi} + \tilde{\mu})$.

<u>$(b_n) \Rightarrow (a_n)$</u>

Let φ^ℓ be a closed form on Δ^n with $\varphi^\ell|\partial\Delta^n = 0$. By the Poincaré lemma, $\varphi^\ell = d\psi^{\ell-1}$ for some form $\psi^{\ell-1}$ which may not vanish on $\partial\Delta^n$. Clearly $\psi^{\ell-1}|\partial\Delta^n$ is closed. If $\ell = 1$ and $n \geq 2$, then $\psi^{\ell-1}|\partial\Delta^n$ is a constant, and hence by subtracting this constant we can make $\psi^{\ell-1}|\partial\Delta^n = 0$. If $\ell > 1$, then by (b_n), $\psi^{\ell-1}|\partial\Delta^n = d\mu^{\ell-2}$. (If $\ell = n$, then $\int_{\partial\Delta^n} \psi^{\ell-1} = \int_{\Delta^n} d\psi$ $= \int_{\Delta^n} \varphi = 0$). Extend μ to a form $\tilde{\mu}$ on all of Δ^n. Then $\varphi = d(\psi - d\tilde{\mu})$ and $(\psi - d\tilde{\mu})|\partial\Delta^n = 0$. ■

This completes the proof of the lemmas. Let us return to the two statements in Proposition 8.1.

(i): <u>Suppose</u> $\varphi \in A^n(K)$ <u>satisfies</u> $d\varphi = 0$ <u>and</u> $\int_{\sigma^n} \varphi = 0$ <u>for all</u> $\sigma^n \in K$. <u>Then</u> $\varphi = d\psi$ <u>for some</u> $\psi \in A^{n-1}(K)$ <u>with the property that</u> $\int_{\tau^{n-1}} \psi = 0$ <u>for all</u> $\tau^{n-1} \in K$.

<u>Proof</u>: Given φ as above there is, according to Lemma 8.4, a form $\psi_\sigma \in A^{n-1}(\sigma^n)$ such that $\varphi|\sigma = d\psi_\sigma$ and such that $\psi_\sigma|\partial\sigma^n = 0$. The collection $\{\psi_{\sigma^n}\}$ defines an element in $A^{n-1}(K^{(n)})$ which vanishes on $K^{(n-1)}$. Using the extension

lemma repeatedly we can extend this form to ψ_n on all of K. Clearly $\varphi - d\psi_n$ vanishes on $K^{(n)}$. The difference $(\varphi - d\psi_n)|\sigma^{n+1}$ is closed and vanishes on $\partial\sigma^{n+1}$. Hence, there is a form $\mu_\sigma \in A^{n-1}(\sigma^{n+1})$ such that $d\mu_\sigma = (\varphi - d\psi_n)|\sigma$ and $\mu_\sigma|\partial\sigma^{n+1} = 0$. As before, the $\{\mu_\sigma\}$ define a form on $K^{(n+1)}$ vanishing on $K^{(n)}$. This can be extended to a form ψ_{n+1} on K which vanishes on $K^{(n)}$. The difference $\varphi - d(\psi_n + \psi_{n+1})$ vanishes on $K^{(n+1)}$. Continuing in this manner we define $\psi_n, \psi_{n+1}, \ldots, \psi_i \in A^{n-1}(K)$, such that $\psi_i|K^{(i-1)} = 0$ and $\varphi - d(\psi_n + \ldots + \psi_{n+k})$ vanishes on $K^{(n+k)}$. Because ψ_i vanishes on $K^{(i-1)}$, the infinite sum $\Sigma_{k=0}^{\infty} \psi_{n+k}$ is an element of $A^*(K)$. It is the required element ψ.

(ii) $A^*(K) \xrightarrow{\rho} C^*(K)$ <u>is onto</u>.

<u>Proof</u>: Given a simplex $\sigma^n \in K$ there is a form in σ^n, $(1/\text{vol}\ \sigma^n)(dt_1 \wedge \ldots \wedge dt_n)$, which has integral 1. This form can be extended by 0 to the rest of the n-skeleton of K. Using the extension lemma we then extend it to all of K. The result is an n-form φ such that

$$\int_{\sigma^n} \varphi = 1 \quad \text{and} \quad \int_{\tau^n} \varphi = 0 \quad \text{for}$$

$$\tau^n \neq \sigma^n.$$

Since we can do this for any simplex, the map ρ is onto.

C. Naturality under subdivision

Let K be a linear cell complex. This means that $K = \cup D_i$ where D_i is a convex linear cell. On the intersection of two cells the linear structures agree. A subdivision of K is a new linear cell structure on K so that each cell in the new decomposition lies linearily in a cell of the old. Such subdivisions arise by choosing points which are to be the new vertices. If we choose exactly one point from each cell, the resulting subdivision is called barycentric subdivision. Barycentric subdivision always yields a simplicial complex.

Lemma 8.5: Let K be a linear cell complex, and let K' be a subdivision of K which is a simplicial complex. Then the integration map

$$\rho : A^*(K') \longrightarrow C^*(K;\mathbb{Q})$$

induces an isomorphism on cohomology.

Proof: The definition of ρ is to integrate forms in $A^*(K')$ on cells of K. Now apply the deRham theorem. ∎

If K is a simplicial complex and K' is a subdivision of K in which each new vertex has rational barycentric coordinates in K, then restriction induces a map of differential algebras,

$$A^*(K) \longrightarrow A^*(K').$$

By the deRham theorem, this map is an isomorphism on cohomology. If $|K| \overset{f}{\to} |L|$ is a continuous map between the geometric realizations of simplicial complexes, then after subdividing K sufficiently to get K', f can be approximated by a simplicial map $\varphi: K' \to L$. This approximation induces $|\varphi|: |K| \to |L|$ which is homotopic to f. Since K' could be any sufficiently fine subdivision of K, we can take it to be a rational subdivision. Hence we will have $\varphi^*: A^*(L) \to A^*(K')$ resulting from a continuous map $f: |K| \to |L|$. Of course φ^* will depend significantly on many choices (i.e.,which subdivision and which approximation we take) but (as we shall see) in the homotopy category of differential algebras φ^* is well defined.

D. <u>Multiplicativity of the de Rham isomorphism</u>.

Let K be a simplicial complex, $A^*(K)$ the P.L. forms on K, and $H^*_{DR}(K)$ its cohomology. We have shown that

$$H^*_{DR}(K) \xrightarrow{\rho^*} H^*(|K|;\mathbb{Q})$$

is an additive isomorphism. Both $H^*_{DR}(K)$ and $H^*(|K|;\mathbb{Q})$ are naturally graded rings, H^*_{DR} from wedge product of forms and $H^*(|K|;\mathbb{Q})$ from the Alexander-Whitney formula for cup product of cochains. We want to prove that

ρ^* <u>is an isomorphism of graded algebras</u>.

The easiest path to this goal is to give a different description of the cup product in singular cohomology. Let K be a simplicial complex with underlying space $|K|$. The space $|K| \times |K|$ carries naturally the structure of a linear cell complex. The cells are all products $\sigma \times \tau$ with σ and

τ simplices of K. This cell structure is not a simplicial complex structure. (If it were, then there would be a graded-commutative, associative cup product on simplicial cochains.) Let α and β be simplicial cochains on K. We define $\alpha \otimes \beta$ as a cellular cochain on $|K| \times |K|$. The value is given by $\langle \alpha \otimes \beta, \sigma \times \tau \rangle = \langle \alpha, \sigma \rangle \cdot \langle \beta, \tau \rangle$. One sees easily that $\delta(\alpha \otimes \beta) = \delta\alpha \otimes \beta + (-1)^{\deg \alpha} \alpha \otimes \delta\beta$. Consequently, if α and β are cocycles, then so is $\alpha \otimes \beta$. The class $[\alpha \otimes \beta] \in H^*(|K| \times |K|)$, and when restricted to the diagonal $\Delta: |K| \to |K| \times |K|$ it gives $[\alpha] \cup [\beta]$ in $H^*(|K|)$.

The Alexander-Whitney formula for multiplication arises from choosing a homotopy from $\Delta: |K| \to |K| \times |K|$ to a linear mapping (called a chain approximation to the diagonal).

Let φ_0 and φ_1 be elements of $A^*(K)$. Let $(|K| \times |K|)'$ be a rational subdivision of $|K| \times |K|$ which is a simplicial complex. Define $\varphi_0 \otimes \varphi_1 \in A^*((|K| \times |K|)')$ as follows. Each simplex $\tau \subset (|K| \times |K|)'$ lies in a product $\sigma_0 \times \sigma_1$ so that its vertices are rational. The form $(\varphi_0|\sigma_0) \otimes (\varphi_1|\sigma_1)$ is a polynomial form with rational coefficients in the product linear structure. Thus $(\varphi_0|\sigma_0) \otimes (\varphi_1|\sigma_1)$ restricts to give a form $\varphi_0 \otimes \varphi_1(\tau) \in A^*(\tau)$. Clearly, these forms fit together to define $\varphi_0 \otimes \varphi_1 \in A^*((|K| \times |K|)')$.

Under the map $\rho: A^*((|K| \times |K|)') \to C^*(|K| \times |K|; \mathbb{Q})$ the form $\varphi_0 \otimes \varphi_1$ goes to the cochain which evaluates on $\sigma_0 \times \sigma_1$ to give $(\int_{\sigma_0} \varphi_0) \cdot (\int_{\sigma_1} \varphi_1)$. Thus, if φ_0 and φ_1 are closed forms, then $\rho(\varphi_0 \otimes \varphi_1)$ is the cocycle $\rho(\varphi_0) \otimes \rho(\varphi_1)$. Restricting to the diagonal we see that the singular cohomology class of $\varphi_0 \wedge \varphi_1$ is the cup product of the classes of φ_0 and φ_1. This proves that $\rho^*: H^*(A^*(K)) \to H^*(C^*(K); \mathbb{Q})$ is an algebra isomorphism.

E. <u>Connection with the C^∞ deRham theorem</u>.

If M is a C^{∞} manifold, then associated to it is the differential algebra of C^{∞} forms. It is an algebra over $\mathbb{R}$. The original theorem of deRham says that the cohomology of this differential algebra is naturally isomorphic (as a ring) to the singular cohomology with real coefficients. The connection between forms on singular cochains is once again achieved by integration. There are many proofs by now of deRham's theorem. For example, one can use currents to give a proof (essentially de Rham's original proof); one can prove that the resulting homology groups satisfy the Eilenberg-Steenroad axioms and hence must be singular homology. More in the spirit of the present discussion one can prove the Poincaré lemma and then establish the isomorphism by induction on a handlebody (instead of a triangulation). Similarly one could use a cover byconvex subsets of $\mathbb{R}^n$, but in this set up the induction is more complicated and goes under the appelation of "sheaf theory."

Whatever method is chosen, the result is the following:[†]

<u>Theorem 8.6</u>: <u>Let</u> $A^*_{C^{\infty}}(M)$ <u>denote the D.G.A of</u> C^{∞} <u>forms on</u> M. <u>Then the a map of cochain complexes induced by integration</u>

$$\rho : A^*_{C^{\infty}}(M) \longrightarrow C^*(M;\mathbb{R})$$

<u>induces an isomorphism of cohomology rings.</u>

Any C^{∞} manifold has a C^{∞} triangulation. This means

[†] See "Geometric Integration Theory" by Whitney, Princeton University Press, 1957.

that there is a simplicial complex K_M and a homeomorphism $\varphi: K_M \to M$ which is C^∞ on each simplex. Consider the D.G.A. of $\mathbb{Q}$-polynomial forms $A^*(K_M)$. This gives a second algebra of forms on M. We wish to relate these. Let

$$A^*(M) = \left\{ \begin{array}{l} \text{P.L. forms on } M \text{ for the} \\ \text{given triangulation} \end{array} \right\}$$

$$A^*_{C^\infty}(M) = \{C^\infty\text{-forms}\}$$

$$\tilde{A}^*(M) = \left\{ \begin{array}{l} \text{collections } \{\varphi\} \text{ of forms on the} \\ \text{simplices } \{\Delta^n\} \text{ such that} \\ \text{(i)} \quad \varphi \quad \text{is } C^\infty \text{ in } \bar{\Delta}^n \text{ for each } n\text{-simplex} \\ \text{(ii)} \quad \varphi|\Delta^n \cap \Delta'^n = \varphi'|\Delta^n \cap \Delta'^n. \end{array} \right\}$$

Thus $\tilde{A}^*(M)$ is constructed in the same way as $A^*(M)$ only now using C^∞ forms. This definition makes sense, and $\tilde{A}^*(M)$ is graded algebra having a "d". Also, Stokes' theorem for simplicial chains holds as before. We call forms in $\tilde{A}^*(M)$ <u>piecewise</u> C^∞<u>-forms</u>. We have:

$$\begin{array}{ccc} A^*(M) \otimes_{\mathbb{Q}} \mathbb{R} & & A^*_{C^\infty}(M) \\ & \searrow \quad \swarrow & \\ & \tilde{A}^*(M) & \end{array}.$$

If we denote the cohomology of $\tilde{A}^*(M)$ by $H^*_{p.C^\infty}(M)$, then the integration map induces a map $H^*_{p.C^\infty}(M) \to H^*(M;\mathbb{R})$. Thus, we have a commutative diagram:

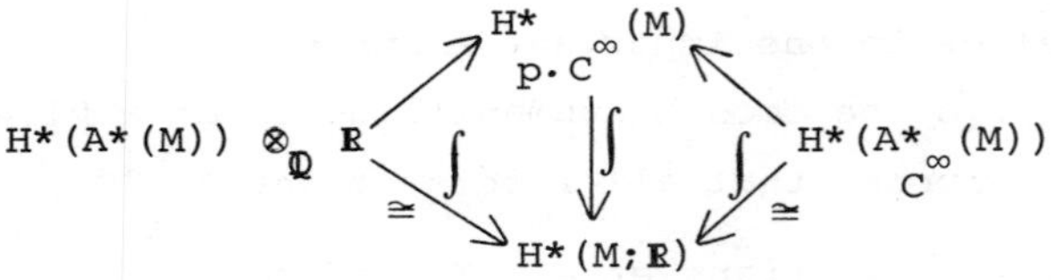

We shall show:

<u>Proposition 8.7</u>: <u>Integration induces an isomorphism</u> $H^*_{p\cdot C^\infty}(M) \to H^*(M;\mathbb{R})$.

<u>Corollary 8.8</u>: <u>The inclusions</u> $A^*(M) \otimes_{\mathbb{Q}} \mathbb{R} \to \widetilde{A}^*(M)$ <u>and</u> $A^*_{C^\infty}(M) \to \widetilde{A}^*(M)$ <u>both induce isomorphisms on cohomology.</u>

<u>Proof of proposition 8.7</u>: The proof is essentially the same as in the PL case. The Poincaré lemma holds and is proved the same way. The extension lemma is again proved using stereographic projection, but instead of multiplying by a power of $(1-t_n)$ we simply multiply by a C^∞ function of t_n which is 1 for $t_n = 0$ and is identically 0 near $t_n = 0$. Once we have these lemmas, the argument given in the $\mathbb{Q}$-polynomial case is valid, mutatis-mutandis, in the piecewise C^∞ case. ■

Corollary 8.8 follows immediately from the commutative diagram and Proposition 8.7.

F. <u>Generalizations of the construction</u>.

Suppose that X is a space made out of manifold "pieces". This includes, for example, a simplicial complex . Then it is possible to define an algebra of piecewise C^∞ forms on X which will calculate the cohomology. We don't prove this result in general, but rather confine ourselves to one important example.

Let $D_1,\ldots,D_k$ be smooth submanifolds in an ambient manifold Y. Suppose that all intersections $D_{i_1} \cap \ldots \cap D_{i_t}$, $i_1 < i_2 < \ldots < i_t$, are transverse. We define $A^*_{p\cdot C^\infty}(\cup_{i=1}^k D_i)$ to be compatible collections of forms

$$\{\omega_i \in A^*_{C^\infty}(D_i),\ i = 1,\ldots,t \text{ satisfying } \omega_i|D_i \cap D_j = \omega_j|D_i \cap D_j \text{ for all } i,j\}.$$

We claim that the cohomology of $A^*_{p\cdot C^\infty}(\cup_{i=1}^k D_i)$ is isomorphic to the singular cohomology of $\cup_{i=1}^k D_i$ with real coefficients. Triangulate the union so that all intersections are subcomplexes and each simplex in the triangulation is C^1. Let K be the resulting simplicial complex. There is a map

$$A^*_{p\cdot C^\infty}(\cup_{i=1}^k D_i) \longrightarrow C^*(K;\mathbb{R})$$

given by integrating the form over the simplices of K. We claim that this map induces an isomorphism on cohomology. This is proved by induction on k. If $k = 1$, then this is exactly the deRham theorem. Suppose we have established the result for all $\ell < k$. We write $\cup_{i=1}^k D_i$ as $(\cup_{i=1}^{k-1} D_i) \cup D_k$ where the intersection is $\cup_{i=1}^{k-1}(D_i \cap D_k)$. Let $D' = \cup_{i=1}^{k-1} D_i$ and $D = \cup_{i=1}^k D_i$. We have a commutative diagram of short exact sequences

$$\begin{array}{ccccccccc} 0 \to & A^*_{p\cdot C^\infty}(D) & \to & A^*_{p\cdot C^\infty}(D') \oplus A^*_{p\cdot C^\infty}(D_k) & \to & A^*_{p\cdot C^\infty}(D' \cap D_k) & \to 0 \\ & \downarrow & & \downarrow \quad\quad \downarrow & & \downarrow & \\ 0 \to & C^*(D;\mathbb{R}) & \to & C^*(D';\mathbb{R}) \oplus C^*(D_k;\mathbb{R}) & \to & C^*(D' \cap D_k;\mathbb{R}) & \to 0 \end{array}$$

The result follows immediately by induction and the 5-lemma.

IX. Differential Graded Algebras.

A. Introduction.

In this section we shall study differential algebras in their own right. What we are doing, actually, is studying the homotopy theory of differential algebras. In fact, we shall construct an object (the minimal model) which should be considered the Postnikov tower of a differential algebra.

Definition 9.1: A differential graded algebra (or differential algebra for short), $\mathcal{A}^*$, is a graded vector space over $\mathbb{Q}$, $\mathbb{R}$ or $\mathbb{C}$,

$$\mathcal{A}^* = \oplus_{p \geq 0} \mathcal{A}^p,$$

having

(i) a differentiation $d: \mathcal{A}^* \to \mathcal{A}^{*+1}$ with $d^2 = 0$;

(ii) a product $\mathcal{A}^p \otimes \mathcal{A}^q \to \mathcal{A}^{p+q}$ satisfying

$$\alpha\beta = (-1)^{pq}\beta\alpha$$

(iii) $d(\alpha\beta) = d\alpha\, \beta + (-1)^p \alpha d\beta$.

We use the notation D.G.A. for a differential graded algebra.

Examples: (i) the C^∞ deRham complex $A^*_{DR}(M)$ of a smooth manifold and the P.L. deRham complex $A^*_{P.L.}(K)$ of a simplicial complex are D.G.A.'s over $\mathbb{R}$, $\mathbb{Q}$ respectively.

(ii) The cohomology $H^*(X,\mathbb{Q})$ of a space is a D.G.A. ($d = 0$), but the singular cochain complex $C^*(X,\mathbb{Q})$ is not (the commutativity fails).

(iii) The problem of commutative cochains. Perhaps the main genesis of the theory we are considering is the problem of commutative cochains. This was solved in an abstract manner by Quillen, and in an attempt to better understand this Sullivan was led to the P.L. forms and the connection between differential forms and homotopy type. In retrospect one can already see much of the theory in the book "Geometric Integration Theory" by Whitney; however one fundamental point was missing in that Whitney only constructs commutative cochains over $\mathbb{R}$, and as already mentioned there is no way to build Postnikov towers over $\mathbb{R}$ and thus tie in the commutative cochains with homotopy type.

Let X be a simplicial complex. The usual definition of the cup-product $\alpha_\alpha \cup \beta_q$ between a p-cochain α_q and a q-cochain β_q is

$$(2) \qquad \langle \alpha_p \cup \beta_q, \Delta^{p+q} \rangle = \langle \alpha_p, \text{ front p-face of } \Delta^{p+q} \rangle \cdot$$

$$\cdot \langle \beta_q, \text{ back q-face of } \Delta^{p+q} \rangle .$$

This formula leads to the properties:

(i) $\delta(\alpha_p \cup \beta_q) = \delta\alpha_p \cup \beta_q + (-1)^p \alpha_p \cup \delta\beta_q$,

and

(ii) $\alpha_p \cup (\beta_q \cup \gamma_r) = (\alpha_p \cup \beta_q) \cup \gamma_r$.

Moreover, a somewhat grizzly computation shows that on the cohomology level we have graded commutativity

$$[\alpha_p] \cup [\beta_q] = (-1)^{pq} [\beta_q] \cup [\alpha_p].$$

However it is very false that

(iii) $\alpha_p \cup \beta_q = (-1)^{pq} \beta_q \cup \alpha_p$

on the cochain level. Now one may attempt to modify the

formula so as to have (i)-(iii), but all such attempts are doomed to failure since, as realized by Steenrod 35 years ago, the failure to find commutative cochains/$\mathbb{Z}$ is reflected in the existence of cohomology operations, such as the Steenrod squares, etc. These objections do not apply over $\mathbb{Q}$ (we have essentially proved this by calculating $H^*(K(\mathbb{Z},n),\mathbb{Q})$), thus it is reasonable to look for commutative cochains/$\mathbb{Q}$. Before going on, it is time to precisely define what is meant by commutative cochains.

<u>Definition</u>: Commutative cochains assign functorially to each simplicial complex X a D.G.A./$\mathbb{Q}$, $C^*(X)$, satisfying (i)-(iii) above and such that

(iv) the cohomology of $\mathcal{C}^*(X)$ is $H^*(X;\mathbb{Q})$; and

(v) given a subcomplex $Y \subset X$, we have

$$C^*(X) \to C^*(Y) \to 0.$$

The problem of commutative cochains is to find such a $\mathcal{C}^*(X)$ for each simplicial complex X.

<u>Example</u>: The P.L. forms $A^*_{PL}(X)$ give an explicit solution to the commutative cochain problem. The cohomology $H^*(X,\mathbb{Q})$ does <u>not</u> give a solution because (v) is violated.

The argument given in this section will also show

<u>Theorem</u>: <u>Let</u> $\mathcal{C}^*(X)$ <u>be any solution to the commutative cochain problem. Then the minimal model</u> $\mathfrak{M}$ <u>of the D.G.A</u>. $C^*(X)$ <u>gives the</u> $\mathbb{Q}$-<u>homotopy type of</u> X.

So now we have some full circle. The problem of commutative cochains is <u>equivalent</u> to finding not only the cohomology, but also the $\mathbb{Q}$-homotopy type of a space from a cochain complex. The P.L. forms explicitly solve this problem, and moreover a simple comparison theorem shows that the C^∞ forms give the $\mathbb{R}$-homotopy type of a smooth manifold.

It is interesting to note that Whitney, in his book, essentially showed that <u>any</u> solution to the commutative cochain problem /$\mathbb{R}$ satisfying a mild continuity condition is given by integration of suitable differential forms (the <u>flat forms</u>) over chains. Now, almost twentyfive years later we have finally understood what he was driving at.

Given a D.G.A., $\mathcal{A}^*$, we denote by $H^*(\mathcal{A}^*)$ the cohomology algebra. It is again a D.G.A. with $d = 0$.

We assume throughout that $H^0(\mathcal{A}^*)$ is the ground field and that $H^1(\mathcal{A}^*) = 0$. Thus, $\mathcal{A}^*$ is, so to speak, <u>connected</u> and <u>simply connected</u>.

<u>Definition</u>: A D.G.A. $\mathcal{A}^*$ is said to be <u>minimal</u> if

(i) $\mathcal{A}^*$ is free as a graded-commutative algebra,

(ii) $\mathcal{A}^1 = 0$, and

(iii) $d(\mathcal{A}^*) \subset \mathcal{A}^+ \wedge \mathcal{A}^+$ where $\mathcal{A}^+ = \oplus_{k>0} \mathcal{A}^k$.

Condition (i) means that $\mathcal{A}^*$ is a tensor product of polynomial algebras on generators of even degrees and exterior algebras on generators of odd degrees. Condition (iii) says that d is <u>decomposable</u>. There is a notion of <u>minimal</u> D.G.A.'s which do not satisfy (ii), (see section XII.)

Given a D.G.A., $\mathcal{A}^*$, we wish to construct a <u>minimal model</u>, $\mathfrak{M}(\mathcal{A}^*)$, for $\mathcal{A}^*$. By definition this means that $\mathfrak{M}(\mathcal{A}^*)$ is a minimal D.G.A. and there is a map $\rho: \mathfrak{M}(\mathcal{A}^*) \to \mathcal{A}^*$ of D.G.A.'s inducing an isomorphism on cohomology. One of the main results of this section is that every simply connected D.G.A. has a minimal model.

<u>Remark</u>: The construction of $\mathfrak{M}(\mathcal{A}^*)$ is motivated by the construction of the Postnikov tower of a space. In fact, the

parallel is quite precise, as we shall see in section XI.

B. Hirsch extensions.

Actually the fundamental property of a minimal algebra is that it is an increasing sequence of subalgebras which are nicely related, one to the next. To illuminate this we study these extensions separately. First

Definition 9.2: Let A be a D.G.A. A Hirsch extension of A is

$$A \to A \otimes_d \Lambda(V)_k.$$

The notation means that i) V is a (finite dimensional) vector space homogeneous of degree k; ii) $\Lambda(V)_k$ is the free graded-commutative algebra with unit generated by V (the polynomial algebra on V if k is even and the exterior algebra on V if k is odd); and iii) $d: V \to A^{k+1}$. The differential on the full algebra is determined by $d|A$ and $d|V$. We write v for $1 \otimes v (v \in V)$ and a for $a \otimes 1$ $(a \in A)$.

Two Hirsch extensions $A \to A \otimes_d \Lambda(V)$ and $A \to A \otimes_d \Lambda(V')$ are equivalent if there is a commutative diagram

$$\begin{array}{ccc} & \nearrow & A \otimes_d \Lambda(V) \\ A & & \downarrow \varphi \\ & \searrow & A \otimes_{d'} \Lambda(V') \end{array}$$

with φ an isomorphism.

Lemma 9.3: $A \to A \otimes_d \Lambda(V)$ and $A \to A \otimes_{d'} \Lambda(V')$ are equivalent if and only if there is an isomorphism $\psi: V \to V'$ so that

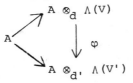

$$\begin{array}{ccc} V & \xrightarrow{d} & H^{k+1}(A) \\ \downarrow{\scriptstyle\psi} & & \downarrow \\ V' & \xrightarrow{d'} & H^{k+1}(A) \end{array}$$

<u>commutes</u>.

<u>Proof</u>: If $\varphi: A \otimes_d \Lambda(V) \xrightarrow{\sim} A \otimes_{d'} \Lambda(V')$ is an isomorphism extending the identity on A, then $\varphi(v) = a_v + \psi(v)$. For φ to be an isomorphism, ψ must be an isomorphism. Since $\varphi(dv) = d'(\varphi(v)) = d'a_v + d'\psi(v)$,

$$[dv] = [d'\psi(v)] \in H^{k+1}(A).$$

Conversely, if we have $\psi: V \xrightarrow{\sim} V'$ so that $[dv] = [d'\psi(v)] \in H^{k+1}(A)$, then $dv - d'\psi(v) = da_v$ for some $a_v \in A$. We can choose a_v linearly in v. Define $\varphi(v) = a_v + \psi(v)$. This defines a map $\varphi: A \otimes \Lambda(V) \to A \otimes \Lambda(V')$, which is easily seen to be an isomorphism and to commute with the differentials.■

To classify Hirsch extensions of A with a fixed vector space of <u>new generators</u> V, we say that two are equivalent if the isomorphism

$$\varphi: A \otimes_d \Lambda(V) \longrightarrow A \otimes_{d'} \Lambda(V)$$

is the identity on A and sends v to $a_v + v$. Equivalence classes are then in a natural one-to-one correspondence with maps

$$d: V \longrightarrow H^{k+1}(A);$$

or what is the same thing, the class of d

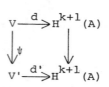

$$[d] \in H^{k+1}(A;V^*).$$

Proposition 9.4: If $\mathfrak{M}$ is a minimal algebra and $\mathfrak{M}(n) \subset \mathfrak{M}$ is the subalgebra generated in degrees $\leq n$, then we have $\mathfrak{M}(0) = \mathfrak{M}(1) \subset \mathfrak{M}(2) \subset \mathfrak{M}(3) \subset \ldots$ with $\cup\, \mathfrak{M}(n) = \mathfrak{M}$, and with each $\mathfrak{M}(n) \subset \mathfrak{M}(n+1)$ being a Hirsch extension.

Proof: Since each $\mathfrak{M}(i)$ is free as an algebra, it is clear that as vector spaces

$$\mathfrak{M}(n+1) \cong \mathfrak{M}(n) \otimes \Lambda(V)_{n+1}.$$

Since $d(v)$ is decomposable for $v \in V_{n+1}$ and $\mathfrak{M}$ has no elements of degree 1, $d(v)$ is a sum of products of elements of degree $\leq n$; i.e., $dv \in \mathfrak{M}(n)$. This proves that $\mathfrak{M}(n) \subset \mathfrak{M}(n+1)$ is a Hirsch extension.

Conversely, if $\mathfrak{M} = \cup_n \mathfrak{M}(n)$ where $\mathfrak{M}(n) \subset \mathfrak{M}(n+1)$ is a Hirsch extension of degree $n+1$ and $\mathfrak{M}(0)$ is the ground field, then $\mathfrak{M}$ is a minimal D.G.A. (as is each $\mathfrak{M}(n)$).■

C. Relative cohomology.

Before beginning the actual construction of $\mathfrak{M}(\mathfrak{A}^*)$ we need a few basic facts about relative cohomology for a map between 2 cochain complexes. Let $C^* \xrightarrow{f} D^*$ denote a degree 0 map (i.e., $f: C^n \to D^n$) between two cochain complexes. Define

$$M_f^n = C^n \oplus D^{n-1}$$

and let $\delta: M_f^n \to M_f^{n+1}$ be given by

$$\begin{pmatrix} \delta_C & 0 \\ f & -\delta_D \end{pmatrix}.$$

One checks easily that $\delta^2 = 0$. We define $H^*(C,D)$ to be $H^*(M_f^*)$. The maps $D^{*-1} \xrightarrow[(-i_2)]{} M_f^*$ and $M_f^* \xrightarrow[\pi_1]{} C^*$ commute with the coboundaries. The resulting maps on cohomology give a long exact sequence

$$\longrightarrow H^n(C) \xrightarrow{f^*} H^n(D) \xrightarrow{\delta} H^{n+1}(C,D) \xrightarrow{i^*} H^{n+1}(C) \xrightarrow{f^*} \cdots .$$

Clearly, this long exact sequence is functorial for commutative diagrams

$$\begin{array}{ccc} C_1 & \xrightarrow{f_1} & D_1 \\ \downarrow & & \downarrow \\ C_2 & \xrightarrow{f_2} & D_2 \end{array} .$$

D. Construction of the minimal model.

Theorem 9.5: If $\mathfrak{A}$ is a differential algebra (simply connected), then $\mathfrak{A}$ has a minimal model.

Proof: We shall construct an increasing sequence of Hirsch extensions

$$\mathfrak{M}(0) \subset \mathfrak{M}(1) \subset \mathfrak{M}(2) \subset \cdots$$

together with maps

$$\rho_n : \mathfrak{M}(n) \longrightarrow \mathfrak{A}$$

so that

$$\rho_n | \mathfrak{M}(k) = \rho_k \quad \text{for} \quad k \leq n.$$

The map $\rho_n : \mathfrak{M}(n) \to \mathfrak{A}$ will be an n-minimal model in the sense tha

(i) $\mathfrak{M}(n)$ is minimal and generated by elements in degrees

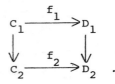

(ii) ρ_n^* is an isomorphism in degrees $\leq n$, and

(iii) ρ_n^* is an injection in degree $n + 1$.

To begin with, $\mathfrak{M}^*(0)$ is the ground field and ρ_0 is the map sending 1 to 1. Suppose inductively that we have constructed $\mathfrak{M}(n)$ and $\rho_n: \mathfrak{M}(n) \to \mathfrak{A}$ as required. The relative cohomology $H^i(\mathfrak{M}(n),\mathfrak{A}))$ vanishes for $i \leq n+1$. (This follows from the long exact sequence of the pair and the conditions on ρ_n.) Let $V = H^{n+2}(\mathfrak{M}(n),\mathfrak{A})$. We will form $\mathfrak{M}(n+1)$ by taking the algebra

$$\mathfrak{M}(n) \otimes \Lambda(V)_{n+1}.$$

Here, $\Lambda(V)_{n+1}$ means the free, graded-commutative algebra generated by the unit together with V in degree n+1. (Hence, $\Lambda(V)_{n+1}$ is the exterior algebra on V if (n+1) is odd and the polynomial algebra on V if n+1 is even.) To extend the differential on $\mathfrak{M}(n+1)$, by the Leibnitz rule we need only define it on V subject to the condition that dv is closed for $v \in V$.

To define a map $\rho_{n+1}: \mathfrak{M}(n+1) \to \mathfrak{A}$ extending ρ_n it is necessary only to define $\rho_{n+1}|V$ subject to the condition

$$\rho_n(dv) = d\rho_{n+1}(v) \quad \text{for all} \quad v \in V.$$

Both d and ρ_{n+1} are defined by splitting the map

$$(\text{Relative cocycles})^{n+2} \underset{s}{\overset{}{\rightleftarrows}} H^{n+2}(\mathfrak{M}(n),\mathfrak{A}).$$

This is equivalent to choosing, linearly in v, cocycle representatives $(m_v,a_v) \in \mathfrak{M}(n)^{n+2} \oplus \mathfrak{A}^{n+1}$. For (m_v,a_v) to be a cocycle means $dm_v = 0$ and $\rho_n(m_v) = da_v$. Having done this we let

$$d(v) = m_v$$

and

$$\rho_{n+1}(v) = a_v.$$

Then

$$d^2(v) = d(m_v) = 0$$

and

$$\rho_n(dv) = \rho_n(m_v) = da_v = d(\rho_{n+1}(v)).$$

This shows that $\mathfrak{M}(n+1)$ is a differential algebra and that ρ_{n+1} is a map of differential algebras. Lastly, we must show that $H^i(\mathfrak{M}(n+1),\mathcal{Q}) = 0$ for $i \leq n+2$. For this we need a lemma.

Lemma 9.6: (a) $\mathfrak{M}(0) = \mathfrak{M}(1)$; i.e., $\mathfrak{M}(n)$ has no elements of degree 1.

(b) $H^{n+2}(\mathfrak{M}(n),\mathfrak{M}(n+1)) = V.$

(c) $H^{n+3}(\mathfrak{M}(n),\mathfrak{M}(n+1)) = 0.$

Proof: (a): Since $\widetilde{H}^0(\mathcal{Q}) = 0$ and $H^1(\mathcal{Q}) = 0$, the relative cohomology $H^2(\mathfrak{M}(0),\mathcal{Q}) = 0$. Hence $\mathfrak{M}(1) = \mathfrak{M}(0)$. But $\mathfrak{M}(n)$ and $\mathfrak{M}(1)$ agree in degree 1 so that $\mathfrak{M}(n)$ has no elements of degree 1.

(b): Let us consider the relative cocycles of degree $(n+2)$. These are all of the form $(a,v+b)$ where $a,b \in \mathfrak{M}(n)$, $v \in V$, $da = 0$, and $a = dv + db$.

Varying such a cocycle by $d(-b,0)$ changes it to (a',v). No element of this form is exact unless $v = 0$.

Conversely, given $v \in V$, we have the cocycle (dv,v). This proves that $H^{n+2}(\mathfrak{M}(n),\mathfrak{M}(n+1)) = V$.

(c) Since $\mathfrak{M}(n)$ has no elements of degree 1, $\mathfrak{M}(n)$ and $\mathfrak{M}(n+1)$ are the same in degree $(n+2)$. It follows easily from this and the fact that $\mathfrak{M}(n) \subset \mathfrak{M}(n+1)$ that

$H^{n+3}(\mathfrak{M}(n),\mathfrak{M}(n+1)) = 0.$■

We have a map of pairs $(id,\rho_{n+1})\colon (\mathfrak{M}_n,\mathfrak{M}_{n+1}) \to (\mathfrak{M}_n,\mathcal{G})$. The corresponding map of long exact sequences, the above fact, and the five lemma prove that $\rho^*_{n+1}\colon H^{n+1}(\mathfrak{M}(n+1)) \to H^{n+1}(\mathcal{G})$ is an isomorphism and that $\rho^*_{n+1}\colon H^{n+2}(\mathfrak{M}(n+1)) \to H^{n+2}(\mathcal{G})$ is an injection.

This completes the inductive construction of the $\mathfrak{M}(n)$ and ρ_n. Define $\mathfrak{M} = \cup_n \mathfrak{M}(n)$ and define $\rho\colon \mathfrak{M} \to \mathcal{G}$ by $\rho|\mathfrak{M}(n) = \rho_n$. Since cohomology commutes with direct limits, it follows that $\rho^*\colon H^*(\mathfrak{M}) \to H^*(\mathcal{G})$ is an isomorphism. Thus, $(\mathfrak{M},\rho)$ is a minimal model for $\mathcal{G}$.■

X. Homotopy Theory of D.G.A.'s

In this section we shall delve more deeply into the homotopy theory of D.G.A.'s. One consequence of this study will be to prove the uniqueness of the minimal model.

A. Homotopies.

Definition 10.1: Let f and g be D.G.A-maps from A to B. A homotopy from f to g is a map

$$H: A \quad B \otimes (t,dt)$$

satisfying $H|_{\substack{t=0\\dt=0}} = f$ and $H|_{\substack{t=1\\dt=0}} = g$.

Explanation: (t,dt) represents the tensor product of polynomials on t (degree of $t = 0$) with the exterior algebra on dt (degree of $dt = 1$). It is thought of as an algebra of forms on the real line, $\mathbb{R}$. The two restrictions correspond to evaluations of forms at the points $\{0\}$ and $\{1\}$. The idea for this defintion comes from dualizing the usual definition on the space level.

To study homotopies we introduce an additive operator

$$\int_0^1 : B \otimes (t,dt) \longrightarrow B$$

by $\int_0^1 b \otimes t^i = 0$ and $\int_0^1 b \otimes t^i dt = (-1)^{\deg b} \frac{b}{i+1}$. Likewise, define

$$\int_0^t : B \otimes (t,dt) \longrightarrow B \otimes (t,dt)$$

by $\int_0^t b \otimes t^i = 0$ and $\int_0^t b \otimes t^i dt = (-1)^{\deg b}\, b \otimes \frac{t^{i+1}}{i+1}$. The following are proved directly from the definitions:

(10.2) If $\beta \in B \otimes (t,dt)$, then

$$d\left(\int_0^t \beta\right) + \int_0^t d\beta = \beta - \{(\beta|_{t=0}) \otimes 1\}.$$

(10.3) If $H: A \to B \otimes (t,dt)$ is a homotopy from f to g, then

$$d\int_0^1 H(a) + \int_0^1 dH(a) = g(a) - f(a).$$

(Note that (10.3) follows from (10.2) by taking $\beta = H(a)$ and restricting to $t = 1$.)

B. Obstruction theory.

The basic result in the homotopy theory of Hirsch extensions is the following:

Proposition 10.4: *Given a diagram*

$$\begin{array}{ccc} A & \xrightarrow{\ g\ } & B \\ \cap\downarrow & & \downarrow\varphi \\ A \otimes_d \Lambda(V)_k & \xrightarrow{\ f\ } & C \end{array}$$

and a homotopy $H: A \to C \otimes (t,dt)$ *from* $\varphi\circ g$ *to* $f|A$, *there is an obstruction class* $\mathcal{O} \in H^{k+1}(B,C;V^*)$ *which vanishes if and only if there is an extension* $\tilde{g}: A \otimes_d \Lambda(V)_k \to B$ *of* g *and an extension* $\tilde{H}$ *of* H *to a homotopy from* f *to* $\varphi\circ\tilde{g}$.

Proof: For each $v \in V$, we define $\tilde{\mathcal{O}}(v) \in B^{k+1} \oplus C^k$ by

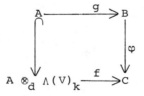

$$\widetilde{\mathcal{O}}(v) = (g(dv), f(v) + \int_0^1 H(dv)). \tag{10.5}$$

This is the obstruction cocycle for extending g and H. First, we show that $\widetilde{\mathcal{O}}(v)$ is indeed a cocycle, and then we show that, if $\widetilde{\mathcal{O}}(v)$ is exact for all $v \in V$, the sought after extensions exist.

$$\begin{aligned} d\widetilde{\mathcal{O}}(v) &= (dg(dv), \varphi \circ g(dv) - df(v) - d\int_0^1 H(dv)) \\ &= (g(d^2v), \widetilde{\varphi} \circ g(dv) - f(dv) - d\int_0^1 H(dv) - \int_0^1 dH(dv)) \\ &= (0,0). \end{aligned}$$

Let $\mathcal{O}: V \to H^{k+1}(B,C)$ be the homomorphism induced by $\widetilde{\mathcal{O}}$; i.e., $\mathcal{O}(v) = [\widetilde{\mathcal{O}}(v)]$. Such a homomorphism is the same as an element $\mathcal{O} \in H^{k+1}(B,C;V^*)$.

If $\mathcal{O}(v) = 0$ for all v, then there are relative cochains (b_v, c_v) (depending linearly on v) so that $d(b_v, c_v) = \mathcal{O}(v)$. Define

$$\widetilde{g}(v) = b_v$$

and

$$\widetilde{H}(v) = f(v) + \int_0^t H(dv) + d(c_v \otimes t).$$

Let us check that these equations define maps of differential algebras $\widetilde{g}: A \otimes_d \Lambda(V)_k \to B$ extending g and $\widetilde{H}: A \otimes_d \Lambda(V)_k \to C \otimes (t,dt)$ extending H. For this it is necessary only that $d\widetilde{g}(v) = g(dv)$ and $d\widetilde{H}(v) = H(dv)$. Clearly, $d\widetilde{g}(v) = db_v = g(dv)$. Also, by (12.2), $d\widetilde{H}(v) = df(v) + H(dv) - H(dv)|_{t=0}$. Since H is a homotopy from f to $\varphi \circ g$, $H(dv)|_{t=0} = f(dv)$. Thus $d\widetilde{H}(v) = H(dv)$. Lastly, we must show that $\widetilde{H}$ is a homotopy from f to $\varphi \circ \widetilde{g}$. But

$$\widetilde{H}(v)|_{t=0} = f(v)$$

and

$$\widetilde{H}(v)|_{t=1} = f(v) + \int_0^1 H(dv) + dc_v$$

$$= f(v) + \int_0^1 H(dv) + (\varphi(b_v) - f(v) - \int_0^1 H(dv))$$

$$= \varphi(b_v) = \varphi\circ\widetilde{g}(v).$$

Conversely, if we are given any extension $\widetilde{g}$ of g and $\widetilde{H}$ of H such that $\widetilde{H}$ is a homotopy from f to $\varphi\circ\widetilde{g}$, then define $\psi_{(\widetilde{g},\widetilde{H})}: V \to B^k \oplus C^{k-1}$ by $\psi(v) = (\widetilde{g}(v), \int_0^1 \widetilde{H}(v))$. One checks directly that $d\psi_{(\widetilde{g},\widetilde{H})} = (g(dv), f(v) + \int_0^1 H(dv)) = \widetilde{\mathcal{O}}(v)$.■

We shall also need a relative version of this lifting property.

<u>Lemma 10.5</u>: <u>Suppose given</u>

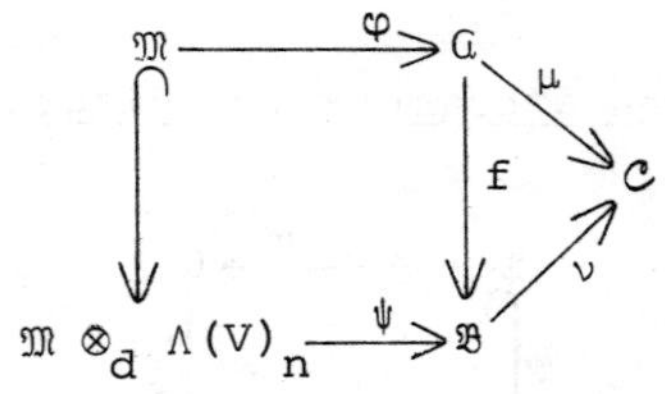

<u>where</u>

(1) $\nu\circ f = \mu$.

(2) μ <u>is onto</u>.

(3) $\mu\circ\varphi = \nu\circ\psi|\mathfrak{M}$

(4) $\mathfrak{M} \xrightarrow{H} \mathfrak{B} \otimes (t,dt) \xrightarrow{\nu\otimes 1} C \otimes (t,dt)$ <u>is constant (i.e.,</u> $(\nu \otimes 1)\cdot(H(m)) = C_m \otimes 1$.

<u>Then, the obstruction cohomology class</u> $\mathcal{O} \in H^{n+1}(\mathfrak{A},\mathfrak{B};V^*)$ <u>vanishes if, and only if, there is an extension</u> $\widetilde{\varphi}$ <u>of</u> φ <u>and an extension</u> $\widetilde{H}: \mathfrak{M} \to \mathfrak{B} \otimes (t,dt)$ <u>of</u> H <u>satisfying</u>

(1) $\mu \circ \tilde{\varphi} = \nu \circ \psi$ <u>and</u>

(2) $(\nu \otimes 1) \circ \tilde{H}$ <u>is a constant homotopy</u>.

<u>Proof</u>: Define $\mathcal{O}: V \to \text{cocycles}^{n+2}(\mathfrak{A}, \mathfrak{B})$ as before:

$$\mathcal{O}(v) = (\tilde{\varphi}(dv), \varphi(v) + \int_0^1 H(dv)).$$

Let $a_v \in \mathfrak{A}$ be such that $\mu(a_v) = \nu(\varphi(v))$. Consider $\mathcal{O}(v) - d(a_v, 0) = (\tilde{\varphi}(dv) - da_v, \varphi(v) - a_v + \int_0^1 H(dv))$. This is a cocycle in $(\ker \mu)^{n+1} \oplus (\text{Ker } \nu)^n$. If it is exact here, $\mathcal{O}(v) - d(a_v, 0) = d(\alpha_v, \beta_v)$ with $\alpha_v \in \text{Ker } \mu$ and $\beta_v \in \text{Ker } \nu$. Then define $\tilde{\varphi}$ and $\tilde{H}$ using the cochain $(a_v + \alpha_v, \beta_v)$. Checking the formulas in 10.4 one sees that $\mu \circ \tilde{\varphi} = \nu \circ \psi$ and that $(\nu \otimes 1) \circ \tilde{H}$ is constant. Since μ is onto, the 5-lemma implies that $H^*(\ker \mu, \text{Ker } \nu) \to H^*(\mathfrak{A}, \mathfrak{B})$ is an isomorphism. Thus $\mathcal{O}(v) - d(a_v, 0)$ is exact in $(\text{Ker } \mu, \text{Ker } \nu)$ if, and only if, $[\mathcal{O}(v)] = 0$ in $H^{n+1}(\mathfrak{A}, \mathfrak{B})$. ■

<u>Corollary 10.6</u>: <u>Given a commutative diagram</u>:

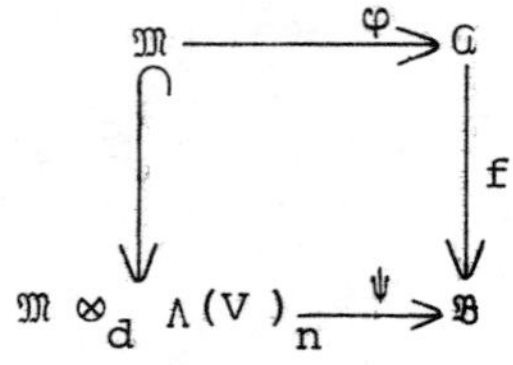

<u>such that</u> f <u>is onto, the element</u> $\mathcal{O}: V \to H^{n+1}(\mathfrak{A}, \mathfrak{B})$ <u>is the obstruction to extending</u> φ <u>to a map</u> $\tilde{\varphi}: \mathfrak{M} \otimes_d \Lambda(V)_n \to \mathfrak{A}$ <u>such that</u> $f \circ \tilde{\varphi} = \psi$.

<u>Proof</u>: Apply 10.5 with $C = \mathfrak{B}$. The cohomology of Ker f is identified with the relative cohomology $H^*(\mathfrak{A}, \mathfrak{B})$. ■

C. Applications of the obstruction theory.

Corollary 10.7: The relation on maps from $\mathfrak{M} \to \mathfrak{A}$, $\mathfrak{M}$ minimal, of being homotopic is an equivalence relation.

Proof: Let $H: \mathfrak{M} \to \mathfrak{A} \otimes (t_1, dt_1)$ be a homotopy from f_0 to f_1 and $J: \mathfrak{M} \to \mathfrak{A} \otimes (t_2, dt_2)$ be a homotopy from f_1 to f_2. Let C be the differential algebra

$$(t_1, t_2, dt_1, dt_2)/\{t_2(t_1 - 1) = 0, t_1 dt_2 = t_2 dt_1 = 0\}.$$

This algebra represents the forms on the variety $t_2(t_1-1) = 0$ in the (t_1, t_2)-plane

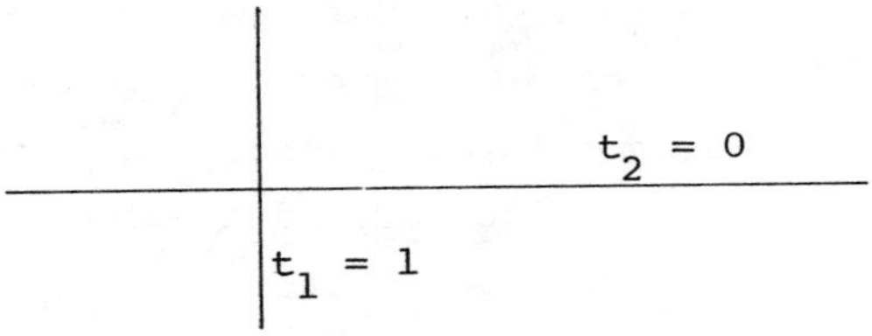

The homotopies H and J define a map "H+J": $\mathfrak{M} \to \mathfrak{A} \otimes \mathcal{C}$. If $H(m) = \Sigma\, a_i \otimes t_1^i + b_i \otimes t_1^i dt_1$ and $J(m) = \Sigma\, a_j' \otimes t_2^j + b_j' \otimes t_2^j dt_2$, then since $H(m)|_{t_1=1} = J(m)|_{t_2=0}$ we have $\Sigma_{i\geq 0}\, a_i = a_0'$. The formula for "H+J"(m) is

$$\Sigma\, a_i \otimes t_1^i + h_i \otimes t_1^i dt + \Sigma_{j\geq 1}\, a_j' \otimes t_2^j + \Sigma_{j\geq 0}\, b_j' \otimes t_2^j dt_2.$$

One checks easily that "H+J" is a map of D.G.A.'s. Consider the diagram

$$\begin{array}{ccc} & & \mathfrak{A} \otimes (t_1, t_2, dt_1, dt_2) \\ & & \big\downarrow p \\ \text{"H+J"}: \mathfrak{M} & \longrightarrow & \mathfrak{A} \otimes \mathcal{C} \end{array}$$

The obstructions to lifting "H+J" lie in
$H^*(\mathcal{G} \otimes [(t_1,t_2,dt_1,dt_2)],\mathcal{C}) = 0$.

Since p is onto, 10.5 says that there is a map $\rho: \mathfrak{M} \to \mathcal{G} \otimes (t_1,t_2,dt_1,dt_2)$ such that $p \circ \rho$ = "H+J". If we restrict ρ to $t_1 = t_2$ we find a homotopy from $f_0 = H|_{t_1=0}$ to $f_2 = J|_{t_2=1}$.

This proves that the relation of being homotopic is transitive. Reflexivity and symmetry are easily shown.■

Theorem 10.8: Let $\varphi: B \to C$ induce an isomorphism on cohomology, and let $\mathfrak{m}$ be a minimal differential algebra. Then $\varphi_*[\mathfrak{M},B] \to [\mathfrak{M},C]$ is a bijection.

Proof: If we have $f: \mathfrak{M} \to C$, then the obstructions to lifting f, up to homotopy, to B step by step over the natural increasing filtration lie in $H^{n+1}(B,C;I^n(\mathfrak{m})^*)$. (Here, $I^n(\mathfrak{m})$ is the space of indecomposables of $\mathfrak{M}$ in degree n. The "*" means the dual vector space.) Since $\varphi: B \to C$ induces an isomorphism on cohomology, $H^*(B,C;V) = 0$ for all vector spaces V. Thus, there is a map $g: \mathfrak{M} \to B$ and a homotopy from f to $\varphi \circ g$. This proves that φ_* is onto.

To show that φ_* is one-to-one, suppose given f_0 and f_1: $\mathfrak{M} \to B$ and a homotopy $H: \mathfrak{M} \to C \otimes (t,dt)$ between $\varphi \circ f_0$ and $\varphi \circ f_1$. Let P be the kernel of:

$$[C \otimes (t,dt)] \oplus B \oplus B \xrightarrow{\mu} C \oplus C \longrightarrow 0,$$

where the map μ sends $\gamma \in C \otimes (t,dt)$ to $(\gamma|_{t=0},\gamma|_{t=1})$ and sends $(b,b') \in B \oplus B$ to $(f(b),f(b'))$. There is a map

$$\rho: B \otimes (t,dt) \longrightarrow P,$$

which sends β to $(f \otimes \mathrm{Id}_{(t,dt)}(\beta),\beta|_{t=0},\beta|_{t=1})$. This map induces an isomorphism on cohomology. To see this notice that one has a long exact sequence

$$\longrightarrow H^*(P) \longrightarrow H^*(C) \oplus H^*(B) \oplus H^*(B) \xrightarrow{\mu^*} H^*(C) \oplus H^*(C) \longrightarrow$$

The map μ^* sends (c,b_0,b_1) to $(c-f^*b_0, c-f^*b_1)$. Hence μ^* is onto and $H^*(P) \subset H^*(C) \oplus H^*(B) \oplus H^*(B)$ is represented as all triples $(c,(f^*)^{-1}(c),(f^*)^{-1}(c))$. From this one sees easily that $\rho: B \otimes (t,dt) \to P$ induces an isomorphism in cohomology.

The maps f_0 and $f_1: \mathfrak{M} \to B$ together with $H: \mathfrak{M} \to (\otimes(t,dt)$ define a map

$$\rho: \mathfrak{M} \longrightarrow P.$$

Apply 10.5 to the diagram

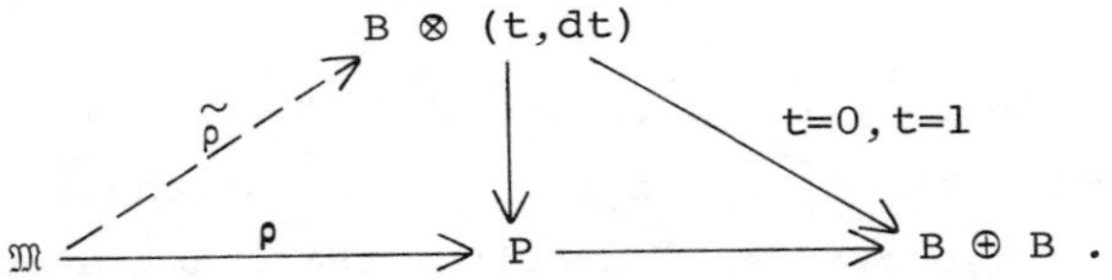

We see that there are no obstructions to lifting ρ to $\tilde{\rho}: \mathfrak{M} \to B \otimes (t,dt)$ such that $\tilde{\rho}|_{i=1} = \pi_{B_i} \circ \rho = f_i$. Thus, f_0 and $f_1: \mathfrak{M} \to B$ are homotopic. ∎

D. <u>Uniqueness of the minimal model.</u>

<u>Theorem 10.9</u>: <u>If</u> $\mathfrak{A}$ <u>is a differential algebra and</u>

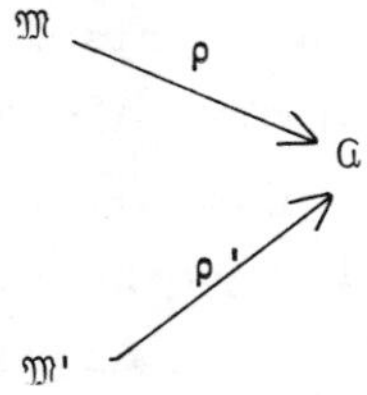

are minimal models for $\mathcal{G}$, *then there is an isomorphism* $I\colon \mathfrak{M} \to \mathfrak{M}'$ *and a homotopy* H *from* ρ *to* $\rho' \circ I$. *The isomorphism* I *is itself determined by these conditions up to homotopy.*

Proof: Applying 10.8 we see that there is a map $I\colon \mathfrak{M} \to \mathfrak{M}'$ satisfying $\rho' \circ I$ is homotopic to ρ, and that such an I is well defined up to homotopy. It remains to show that any such $I\colon \mathfrak{M} \to \mathfrak{M}'$ is an isomorphism. Since, on the level of cohomology, $I^* \circ (\rho')^* = \rho^*$ and $(\rho')^*$ and ρ^* are isomorphisms, it follows that any such $I\colon \mathfrak{M} \to \mathfrak{M}'$ will induce an isomorphism on cohomology. To conclude the proof we have the

Lemma 10.10: *If* $I\colon \mathfrak{M} \to \mathfrak{M}'$ *induces an isomorphism on cohomology (with* $\mathfrak{M}$ *and* $\mathfrak{M}'$ *minmal), then* I *is an isomorphism.*

Proof: Clearly I induces $I_n\colon \mathfrak{M}(n) \to \mathfrak{M}'(n)$. Assuming inductively that I_n is an isomorphism we shall show that I_{n+1} is an isomorphism. Consider the exact sequence of cohomology:

$$
\begin{array}{ccccccccccc}
\to H^{n+1}(\mathfrak{M}(n)) & \to & H^{n+1}(\mathfrak{M}) & \to & H^{n+2}(\mathfrak{M}(n),\mathfrak{M}) & \to & H^{n+2}(\mathfrak{M}(n)) & \to & H^{n+2}(\mathfrak{M}) \\
\downarrow I_n^* & & \downarrow I^* & & \downarrow (I_n, I)^* & & \downarrow I_n^* & & \downarrow I^* \\
H^{n+1}(\mathfrak{M}'(n)) & \to & H^{n+1}(\mathfrak{M}') & \to & H^{n+2}(\mathfrak{M}'(n),\mathfrak{M}') & \to & H^{n+2}(\mathfrak{M}'(n)) & \to & H^{n+2}(\mathfrak{M}')
\end{array}
$$

By assumption the 2nd and 5th arrows are isomorphisms. By the inductive hypothesis so are the 1st and 4th. Hence, the middle map is also an isomorphism.

We claim that $H^{n+2}(\mathfrak{M}(n),\mathfrak{M}) = H^{n+2}(\mathfrak{M}(n),\mathfrak{M}(n+1))$. As we have seen before, $H^{n+2}(\mathfrak{M}(n),\mathfrak{M}(n+1)) \cong V_{n+1}$, the vector space of new generators added in going from $\mathfrak{M}(n)$ to $\mathfrak{M}(n+1)$. To prove the claim note that in degree $\leq n+2$ the relative cochains for $(\mathfrak{M}(n),\mathfrak{M}(n+1))$ and $(\mathfrak{M}(n),\mathfrak{M})$ are equal. In degrees $> n+2$ the former are a subgroup of the latter. The equality of the relative cohomology groups up through degree $(n+2)$ follows immediately from this.

Since these identifications are natural we have a commutative diagram:

$$\begin{array}{ccc}
H^{n+2}(\mathfrak{M}(n),\mathfrak{M}) & \xrightarrow[\cong]{(I_n,I)^*} & H^{n+2}(\mathfrak{M}'(n),\mathfrak{M}') \\
\uparrow \cong & & \uparrow \cong \\
V_{n+1}(\mathfrak{M}) & \xrightarrow{I} & V_{n+1}(\mathfrak{M}').
\end{array}$$

Thus I induces an isomorphism on the vector space of new generators, and hence I_{n+1} induces an isomorphism:

$$\mathfrak{M}(n+1) \xrightarrow{I_{n+1}} \mathfrak{M}'(n+1). \qquad \blacksquare$$

Corollary 10.11: *Let* $\rho_A\colon \mathfrak{M}_A \to A$ *and* $\rho_B\colon \mathfrak{M}_B \to B$ *be minimal models, and let* $f\colon A \to B$ *be a map of differential algebras. There is a map* $\hat{f}\colon \mathfrak{M}_A \to \mathfrak{M}_B$ *and a homotopy from* $\rho_B \circ \hat{f}$ *to* $\hat{f}\circ\rho_A$. *The map* $\hat{f}$ *is determined up to homotopy by these properties.*

Proof: This is immediate from 10.8 applied to the following diagram:

$$\begin{array}{ccc}
 & & \mathfrak{M}_B \\
 & \nearrow & \uparrow \rho_B \\
f\circ\rho_A\colon \mathfrak{M}_A & \longrightarrow & B
\end{array}$$

since ρ_B induces an isomorphism on cohomology. $\blacksquare$

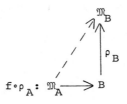

<u>Example</u>: One note of caution is necessary here. On the object level a minimal model is uniquely determined. This is <u>not</u> true on the map level. Here is an example of a nonzero map $\mathfrak{N}^* \xrightarrow{\rho} \mathfrak{M}^*$ between minimal D.G.A.'s/$\mathbb{Q}$ which is homotopic to a constant. Let

$$\mathfrak{N}^* = \{\alpha^2, \beta^3, \gamma^4 : d\alpha = 0 = d\beta, d\gamma = \alpha\cdot\beta\}$$

$$\mathfrak{M}^* = \{\omega^2, \eta^3 : d\eta = \omega\cdot\omega\}$$

$$\rho(\alpha) = \rho(\beta) = 0,\ \rho(\gamma) = \omega\cdot\omega$$

This map is homotopic to zero. One way to see this is to give explicitly the homotopy. Anticipating the connection between D.G.A's over $\mathbb{Q}$ and rational homotopy theory, one can give a geometric argument. On the level of spaces we have $X = S^2$ which is represented by $\mathfrak{M}^*$ and Y given by

$$\begin{array}{ccc} K(\mathbb{Q},4) & \longrightarrow & Y \\ & & \downarrow \pi \\ & & K(\mathbb{Q},2)\times K(\mathbb{Q},3). \end{array}$$

The map ρ is realized by a map $f\colon X \to Y$. By the Whitehead theorem, $\pi\circ f \sim$ constant, so that f is homotopic to a map of X into a fiber. This map $X \xrightarrow{f} K(\mathbb{Q},4)$ is given by the class $\rho(\gamma) = \omega\cdot\omega = 0$ in $H^4(X,\mathbb{Q})$, and so $f \sim$ constant.

XI. The Connection between the Homotopy Theory of D.G.A.'s and Rational Homotopy Theory

The first 7 sections dealt with classical homotopy theory culminating in the study of the rational Postnikov tower of a (simply connected) space. Section VIII was devoted to defining a suitable algebra of differential forms on a simplicial complex. In sections IX and X we studied the homotopy theory of D.G.A.'s in general. In this section we shall bring together these two theories and explain how the rational homotopy theory of a simplicial complex is determined by the minimal model of its p.l. forms. In fact we shall see that $\mathbb{Q}$-Postnikov towers and $\mathbb{Q}$ minimal models are perfectly dual.

A. Transgression in the Serre spectral sequence and the duality.

The basis for the duality alluded to above is the duality between principal fibrations and Hirsch extensions. To understand this we study the transgression in the Serre spectral sequence. Let $p: E \to B$ be a principal fibration with fiber $K(\pi,n)$. Let $\{E_r^{p,q}, d_r\}$ be the Serre spectral sequence of this fibration with coefficients π. Then $E_2^{p,q} = 0$ for $0 < q < n$ and $E_2^{0,n} = H^n(K(\pi,n),\pi) = \mathrm{Hom}(\pi,\pi)$. As a result $E_r^{0,n} = E_2^{0,n}$ for $r \leq n$. The first nonzero differential on $E_*^{0,n}$ is $d_{n+1}: E_n^{0,n} \to E_n^{n+1,0} = H^{n+1}(B;\pi)$. This map $d_{n+1}: \mathrm{Hom}(\pi,\pi) \to H^{n+1}(B;\pi)$ is the transgression map. The class $d_{n+1}(\iota) \in H^{n+1}(B;\pi)$ is the k-invariant of the fibration. (Here ι is the class corresponding to the identity homomorphism of π to π.) If we take $\mathbb{Q}$-coefficients instead of π coefficients, then $d_{n+1}: \mathrm{Hom}(\pi,\mathbb{Q}) \to H^{n+1}(B;\mathbb{Q})$. This map is dual to an element $[d_{n+1}] \in H^{n+1}(B;\pi \otimes \mathbb{Q})$. If π itself is a rational vector space (of finite dimension), then $[d_{n+1}] \in H^{n+1}(B;\pi \otimes \mathbb{Q}) = H^{n+1}(B;\pi)$ is again the k-invariant of the fibration. Thus a principal fibration where the fiber is a local Eilenberg-MacLane space (and the group is of finite dimension over $\mathbb{Q}$) is completely determined by the homotopy group π and $[d_{n+1}] \in H^{n+1}(B;\pi)$.

On the other hand given a $\mathbb{Q}$-vector space π (of finite dimension) and an element $[d] \in H^{n+1}(B;\pi)$ there is a Hirsch extension

$$A^*(B) \otimes_d \Lambda(\pi^*)_n$$

where $d: \pi^* \to$ closed forms of A^{n+1} induces $[d] \in H^{n+1}(B;\pi)$. By Lemma 9.3, this Hirsch extension is well defined up to equivalence. Clearly then, there is a bijection correspondence between principal fibrations over B with fiber a local Eilenberg-MacLane space and Hirsch extensions of $A^*(B)$.

Let $p: \mathcal{E} \to B$ be a fibration with B a simplicial complex. A simplicial model for p is a map $f: \mathcal{E} \to B$ such that

(1) E is a simplicial complex,

(2) $f: E \to B$ is a simplicial map, and

(3) there is a map $i: E \to \mathcal{E}$ such that $p \circ i = f$.

Theorem 11.1: Let $p: \mathcal{E} \to B$ be a principal fibration with fiber $K(V,n)$, V a $\mathbb{Q}$-vector space of finite dimension. Let $f: \mathcal{E} \to B$ be a simplicial model for p. Let $A^*(B) \otimes_d \Lambda(V^*)_n$ be the corresponding Hirsch extension to p. There is a map

$$\rho: A^*(B) \otimes_d \Lambda(V^*)_n \longrightarrow A^*(E)$$

such that:

(1) $\rho | A^*(B)$ is f^*, and

(2) ρ induces an isomorphism on cohomology.

The given Hirsch extension is the only one, up to isomorphism, that admits such a mapping ρ.

This is the connection between principal $K(\pi,n)$ fibrations and Hirsch extensions. The theorem is proved using the Serre spectral sequence by what is, at the core, a simple argument. There are, however, enormous technical problems due to the fact that $f: E \to B$ is not a fibration. In fact simplicial maps and fibrations are rather incompatible. Because of these technical problems we postpone the proof of this theorem to appendix A1.

There is also a converse which is easily deduced from this result.

Theorem 11.2: Let $A^*(B) \otimes_d \Lambda(V)_n$ be a Hirsch extension with V a $\mathbb{Q}$-vector space of finite dimension. There is a principal fibration $p: \mathcal{E} \to B$ with fiber $K(V^*,n)$, and a simplicial model for p, $f: E \to B$ so that there is a map of D.G.A.'s

$$\rho : A^*(B) \otimes_d \Lambda(V)_n \longrightarrow A^*(E)$$

which extends f^* on $A^*(B)$ and induces an isomorphism on cohomology. Up to equivalence there is only one such principal fibration with local fiber.

Proof: We have the identification of $H^*(A^*(B))$ with $\mathcal{H}^*(B;\mathbb{Q})$. The map $V \overset{d}{\to} H^{n+1}(B;\mathbb{Q})$ determines an element $k \in H^{n+1}(B;V^*)$ which in turn determines a principal fibration

$$\begin{array}{ccc} K(V^*,n) & \longrightarrow & \mathcal{E} \\ & & \downarrow \\ & & B \end{array}$$

The transgression in the Serre spectral sequence of this fibration

$$\begin{array}{ccc} \mathrm{Hom}(V^*,\mathbb{Q}) & \longrightarrow & H^{n+1}(B;\mathbb{Q}) \\ \cong\uparrow & & \\ V & & \end{array}$$

is exactly $d\colon V \to H^{n+1}(B;\mathbb{Q})$. According to 11.1, if $E \xrightarrow{f} B$ is a simplicial model for $\mathcal{E} \to B$ then there is a Hirsch extension

$$A^*(B) \otimes \Lambda(W)_n \xrightarrow{\rho'} A^*(E)$$

extending f^* on $A^*(B)$ and inducing an isomorphism in cohomology. We wish to know that this extension is isomorphic to the one with which we began. Clearly, $W \cong H^{n+1}(B,\mathcal{E};\mathbb{Q}) = H^{n+1}(B,\mathcal{E};\mathbb{Q}) = H^n(\text{fiber};\mathbb{Q})$ and $d\colon W \to H^{n+1}(B;\mathbb{Q})$ is exactly the transgression in the spectral sequence of the fibration. But by construction $V = H^n(\text{fiber};\mathbb{Q})$ and the transgression is $d\colon V \to H^{n+1}(B;\mathbb{Q})$. This proves that there is a map

$$\rho\colon A^*(B) \otimes_d \Lambda(V) \longrightarrow A^*(E)$$

extending f^* and inducing an isomorphism on cohomology. ∎

Let B be a simplicial complex. Consider the set of equivalence classes of all principal fibrations over B with fiber an Eilenberg-MacLane space such that $\mathrm{Hom}(\pi_*(\text{fiber}),\mathbb{Q})$ is a finite dimensional rational vector space. Denote this set by $\mathcal{PF}(B)$. If, in addition, we require the homotopy group of the fiber itself to be a (finite dimensional) rational vector space, then denote the subset by $\mathcal{PF}_{\mathbb{Q}}(B)$. Denote the set of equivalence classes of finite dimensional Hirsch extensions of $A^*(B)$ by $\mathcal{HE}(B)$. These three are homotopy functors of B. The above equivalences give natural transformations

$$\mathcal{H}: \mathcal{PF}(B) \longrightarrow \mathcal{HE}(B)$$

and

$$\mathcal{F}: \mathcal{HE}(B) \longrightarrow \mathcal{PF}_{\mathbb{Q}}(B).$$

The composition $\mathcal{F} \circ \mathcal{H}: \mathcal{PF}(B) \to \mathcal{PF}_{\mathbb{Q}}(B)$ is simply localization. The composition

$$\mathcal{H} \circ \mathcal{F}: \quad \mathcal{E}(B) \longrightarrow \mathcal{HE}(B)$$

is the identity. Thus, $\mathcal{F}$ and $\mathcal{H} | \mathcal{PF}_{\mathbb{Q}}(B)$ are inverse functors.

Under these correspondences the degree of the extension is the dimension of the nontrivial homotopy group in the fiber. The vector space of the extension and the homotopy group of the fiber are dual vector spaces. The k-invariant of the fibration is the map induced by d.

<u>Lemma 11.3</u>: <u>Suppose</u> $f: \mathfrak{A} \to \mathfrak{B}$ <u>induces an isomorphism on cohomology. Then</u> f <u>induces a bijection</u>

$$\cdot\mathcal{H}\cdot\mathcal{E}(\mathfrak{A}) \xrightarrow{f^*} \mathcal{E}(\mathfrak{B}).$$

<u>Corollary 11.4</u>: <u>If</u> $\mathfrak{M} \to A^*(B)$ <u>is a minimal model, then we have a bijection</u>

$$\cdot\mathcal{H}\cdot\mathcal{E}(\mathfrak{M}) \longrightarrow \mathcal{PF}_{\mathbb{Q}}(B).$$

B. $\mathbb{Q}$-<u>Postnikov towers and minimal D.G.A.'s</u>

Having established the equivalence of Hirsch extensions with principal fibrations, we turn to the connection between $\mathbb{Q}$-Postnikov towers and minimal D.G.A.'s.

Let X be a simplicial complex which is simply connected. Let

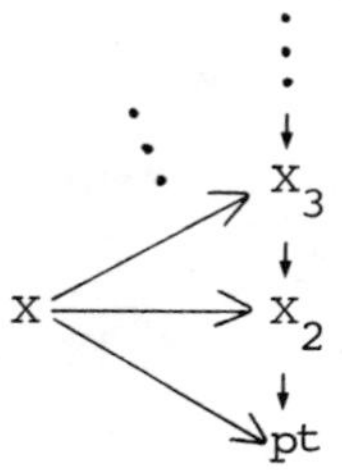

be the rational Postnikov tower of X. Let

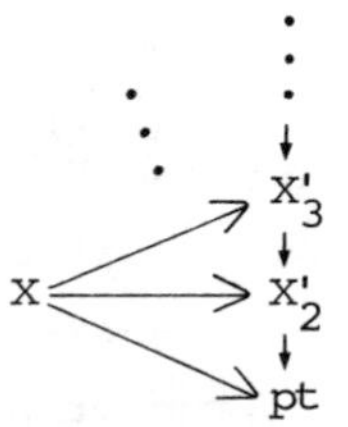

be a simplicial model for this tower of fibrations. Let $\mathfrak{M}(0) \subset \mathfrak{M}(2) \subset \mathfrak{M}(3) \subset \ldots \cup \mathfrak{M}(n) = \mathfrak{M}$ be the minimal model for X.

Theorem 11.5: $\mathfrak{M}(n)$ is the minimal model for X'_n. The Hirsch extension $\mathfrak{M}(n) \subset \mathfrak{M}(n+1)$ corresponds to the principal fibration $X_{n+1} \to X_n$. Thus, $H^*(\mathfrak{M}(n)) = H^*(X_n;\mathbb{Q})$. The indecomposables of $\mathfrak{M}(n+1)$ in degree $(n+1)$, I^{n+1}, are equal to $\mathrm{Hom}(\pi_{n+1}(X),\mathbb{Q})$ and the k-invariant $k^{n+2} \in H^{n+2}(X_n;\pi_{n+1})$, when tensored with $\mathbb{Q}$, is equal to $d\colon I^{n+1} \to H^{n+2}(\mathfrak{M}(n))$.

Proof: We build, inductively, minimal models for the $A^*(X'_n)$. Suppose $\mathfrak{M}(n) \to A^*(X'_n)$ is a minimal model. Then, the minimal model for $A^*(X'_{n+1})$ is a Hirsch extension $\mathfrak{M}(n) \otimes_d \Lambda(V)_{n+1}$ where $V^* = \pi_{n+1}(X) \otimes \mathbb{Q}$ and $d\colon V \to H^{n+2}(M(n))$

corresponds to the k-invariant. Thus, we have commutative diagrams:

$$A^*(X'_{n+1}) \xrightarrow{\rho_{n+1}} \mathfrak{M}(n+1)$$

$$A^*(X'_n) \xrightarrow{\rho_n} \mathfrak{M}(n)$$

where $\rho_k: \mathfrak{M}(k) \to A^*(X'_k)$ is a minimal model. Let $\mathfrak{M} = \cup_{n\geq 0}\mathfrak{M}(n)$. It is a minimal D.G.A. which corresponds to the rational Postnikov tower of X as described in the theorem. To complete the proof we must construct a map $\rho: \mathfrak{M} \to A^*(X)$ inducing an isomorphism on cohomology.

For each n we have a map $f_n: X \to X'_n$. There is a rational subdivision $S_n(X)$ so that f_n is homotopic to a simplicial map, g_n. In fact, we can choose $S_n(X)$ to be a rational subdivision of $S_{n-1}(X)$. Let $A^*{}_n(X)$ be the $\mathbb{Q}$-polynomial forms in $S_n(X)$. We have

$$\begin{array}{ccccc} \cdots \hookrightarrow & A^*_n(X) & \hookrightarrow & A^*_{n+1}(X) & \hookrightarrow \cdots \\ & g^*_n & & g^*_{n+1} & \\ \to & A^*(X'_n) & \longrightarrow & A^*(X'_{n+1}) & \to \cdots \end{array}$$

Each square is homotopy commutative and $A^*_n(X) \hookrightarrow A^*_{n+1}(X)$ induces an isomorphism on cohomology. Let $A^*_{p.l.}(X) = \lim A^*_n(X)$. Clearly $A^*(X) \hookrightarrow A^*_{p.l.}(X)$ induces an isomorphism on cohomology.

The maps $g^*_n \circ \rho_n: \mathfrak{M}_n \to A^*_n(X) \subset A^*_{p.l.}(X)$ are n-minimal models. Unfortunately, there is no reason to expect that $(g_{n+1})^* \circ \rho_{n+1}|\mathfrak{M}_n = g^*_n \circ \rho_n$. On the other hand $(g_{n+1})^* \circ \rho_{n+1}|\mathfrak{M}_n$ is homotopic to $g^*_n \circ \rho_n$ since the above diagram is homotopy commutative.

Let $H: \mathfrak{M}(n) \to A^*_{p.l.}(X) \otimes (t,dt)$ be a homotopy from $g^*_n \circ \rho_n$ (at $t = 0$) to $g^*_{n+1} \circ \rho_{n+1}$ (at $t = 1$). There is a commutative diagram

$$\begin{array}{ccc} \mathfrak{M}(n) & \xrightarrow{\quad H \quad} & A^*_{p.l.}(X) \otimes (t,dt) \\ \cap\big\downarrow & & \big\downarrow t=1 \\ \mathfrak{M}(n+1) & \xrightarrow{\; g^*_{n+1} \circ \rho_{n+1} \;} & A^*_{p.l.}(X). \end{array}$$

Since $H^*(A^*_{p.l.}(X) \otimes (t,dt)) \xrightarrow{t=1} H^*(A^*_{p.l.}(X))$ is an isomorphism, there is no obstruction to extending H over all of $\mathfrak{M}(n+1)$. Since the restriction to $t = 1$ is onto, this extension can be chosen so that the extended homotopy at $t = 1$ is $g^*_{n+1} \circ \rho_{n+1}$. This means that $g^*_{n+1} \circ \rho_{n+1}$ is homotopic to a map $\varphi_{n+1}: \mathfrak{M}(n+1) \to A^*_{p.l.}(X)$ so that $\varphi_{n+1}|\mathfrak{M}(n) = g^*_n \circ \rho_n$.

This allows us to deform the $g^*_n \circ \rho_n: \mathfrak{M}(x) \to A^*_{p.l.}(X)$, by induction on n, to maps $\varphi_n: \mathfrak{M}(n) \to A^*_{p.l.}(X)$ so that $\varphi_{n+1}|\mathfrak{M}(n) = \varphi_n$. The $\{\varphi_n\}$ define a map $\varphi: \mathfrak{M} \to A^*_{p.l.}(X)$. This map is an n-equivalence for all $n \geq 0$. Thus φ induces an isomorphism on cohomology, and hence $\mathfrak{M}$ is the minimal model for $A^*_{p.l.}(X)$. Since $A^*(X) \subset A^*_{p.l.}(X)$ induces an isomorphism on cohomology, $\mathfrak{M}$ is also the minimal model of $A^*(X)$.∎

This theorem tells us explicitly how to construct the minimal model of $A^*(X)$ from the rational Postnikov tower of X and vice-versa. Thus, all rational homotopy theoretic information about X is contained in the minimal model of the forms $A^*(X)$.

<u>Corollary 11.6</u>: <u>The minimal model of the C^∞ forms on a simply connected C^∞ manifold</u> M, $\mathfrak{M}_{C^\infty}(M)$, <u>is isomorphic to</u> $\mathfrak{M}_M \otimes_{\mathbb{Q}} \mathbb{R}$ <u>where</u> $\mathfrak{M}_M$ <u>is dual (in the sense of Theorem 11.5) to the rational Postnikov tower of</u> M.

Proof: Choose a C^∞ triangulation of M. This gives a diagram:

$$A^*(M) \otimes_{\mathbb{Q}} \mathbb{R} \longrightarrow \tilde{A}^*(M) \longleftarrow A^*_{C^\infty}(M)$$

where both inclusions induce isomorphisms on cohomology. Thus, the minimal models of $A^*_{C^\infty}(M)$ and $\tilde{A}^*(M)$ are isomorphic, and the minimal model of $A^*(M)$ tensored with $\mathbb{R}$ is isomorphic to that of $\tilde{A}^*(M)$. Hence, we have an isomorphism, well-defined up to homotopy

$$\mathfrak{M}(A^*(M)) \otimes_{\mathbb{Q}} \mathbb{R} \cong \mathfrak{M}(A^*_{C^\infty}(M)).$$

The corollary now follows immediately from 11.5. ■

Corollary 11.7: Let Y be a smooth manifold and $D_1, \ldots, D_k \subset Y$ smooth submanifolds which intersect transversally. Let $D = \cup_{i=1}^{k} D_i$, and let $A^*_{p.C^\infty}(D)$ be the D.G.A. constructed in Section VIII F. Suppose that D is simply connected. The minimal model of $A^*_{p.C^\infty}(D)$ is the real form of a rational minimal model which is dual to the rational Postnikov tower of D.

Proof: Let K be a C^∞-triangulation of D. There is an inclusion map $A^*_{p.C^\infty}(D) \longrightarrow A^*_{p.C^\infty}(K)$ which gives rise to a commutative diagram of cochain complexes

$$\begin{array}{ccc} A^*_{p.C^\infty}(D) & \longrightarrow & A^*_{p.C^\infty}(K) \\ & \searrow & \downarrow \\ & & C^*(K;\mathbb{R}) \end{array} .$$

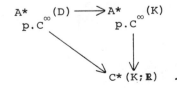

By the result in VIII F the integration map induces an isomorphism in cohomology $H^*(A^*_{p.C^\infty}(D)) \stackrel{\sim}{\to} H^*(K;\mathbb{R})$. By the piecewise-$C^\infty$ deRham theorem, integration induces an isomorphism $H^*(A^*_{p.C^\infty}(D)) \stackrel{\sim}{\to} H^*(K;\mathbb{R})$. Thus the inclusion $A^*_{p.C^\infty}(D) \quad A^*_{p.C^\infty}(K)$ is a homotopy equivalence of D.G.A.'s. The corollary now follows from Theorem 11.5.■

XII. The Fundamental Group

So far we have been considering simply connected spaces only. In this section we broaden the definition to include information about the fundamental group. We find that the theory developed in the previous section has a natural analogue which relates differential forms to the nilpotent part of the fundamental group.

A. 1-minimal Models.

Let $\mathfrak{A}$ be a connected, but not necessarily simply connected, differential algebra. A *1-minimal model* for $\mathfrak{A}$ is a map

$$\rho : \mathfrak{M}_1 \longrightarrow \mathfrak{A}$$

inducing an isomorphism on H^1 and an injection on H^2 where $\mathfrak{M}_1$ is an increasing union of Hirsch extensions of degree 1:

$$\text{Ground Field} = \mathfrak{M}_{1,0} \subset \mathfrak{M}_{1,1} \subset \mathfrak{M}_{1,2} \subset \dots .$$

$$\cup_{n=0}^{\infty} \mathfrak{M}_{1,n} = \mathfrak{M}_1 .$$

Theorem 12.1: Any connected differential algebra has a 1-minimal model. Given two such for $\mathfrak{A}$

$$\begin{array}{ccc} \mathfrak{M}_1 & \searrow^{\rho} & \\ & & \mathfrak{A} \\ \mathfrak{M}_1' & \nearrow_{\rho'} & \end{array} ,$$

there is an isomorphism $I : \mathfrak{M}_1 \rightarrow \mathfrak{M}_1'$ *and a homotopy* H *from*

ρ <u>to</u> $\rho'\cdot I$.

<u>Proof</u>: Let $\mathfrak{M}_{1,1} = \Lambda(H^1(\mathfrak{a}))$ and let $\rho_1: \mathfrak{M}_{1,1} \to \mathfrak{a}$ be defined by sending each cohomology class to a closed form representing it. Given $\rho_n: \mathfrak{m}_{1,n} \to A$ with ρ_n inducing an isomorphism on H^1, define V_{n+1} to be Ker $\rho_n: H^2(M_{1,n}) \to H^2(\mathfrak{a})$. (This is also $H^2(\mathfrak{M}_{1,n},\mathfrak{a})$.)

We choose a splitting s: $H^2(\mathfrak{M}_{1,n},\mathfrak{a}) \to \text{Cocycles}^2(\mathfrak{M}_{1,n},\mathfrak{a})$ for the natural quotient map: $\text{Cocycles}^2(\mathfrak{M}_{1,n},\mathfrak{a}) \to H^2(\mathfrak{M}_{1,n},\mathfrak{a})$, $s(v) = (m_v, a_v)$. Define d: $V_{n+1} \to (\mathfrak{M}_{1,n})^2$ by $d(v) = m_v$; define $\rho: V_{n+1} \to \mathfrak{a}^1$ by $\rho(v) = a_v$. As before one sees that these formulae lead to $\mathfrak{M}_{1,n+1} = \mathfrak{M}_{1,n} \otimes_d \Lambda(V_{n+1})$ and $\rho_{n+1}: \mathfrak{M}_{1,n+1} \to \mathfrak{a}$ extending ρ_n. One sees that
$\text{Ker}(\rho_n^*: H^2(\mathfrak{M}_{1,n}) \to H^2(\mathfrak{a}))$ is contained in
$\text{Ker}(\iota_n^*: H^2(\mathfrak{M}_{1,n}) \to H^2(\mathfrak{M}_{1,n+1}))$ where ι_n is the inclusion $\mathfrak{M}_{1,n} \subset \mathfrak{M}_{1,n+1}$. Thus, in creating $\mathfrak{M}_{1,n+1}$ we have killed the kernel of ρ_n on H^2. But in doing so we may have created a new nonzero kernel at the next stage. We keep repeating the process until $\text{Ker}(\rho_K^*: H^2(\mathfrak{M}_{1,K}) \to H^2(\mathfrak{a}))$ is trivial. If this never happens, then we construct $\mathfrak{M}_{1,n}$ for all $1 \leq n < \infty$, and let $\mathfrak{M}_1 = \cup_n \mathfrak{M}_{1,n}$ with $\rho: \mathfrak{M}_1 \to \mathfrak{a}$ being defined by $\rho|\mathfrak{M}_{1,n} = \rho_n$. One sees easily, from the fact that, on H^2, Ker $\rho_n^* \subset$ Ker ι_n^*, that $\text{Ker}(\rho^*: H^2(\mathfrak{M}) \to H^2(\mathfrak{a})) = 0$. On the other hand, since $\rho_n^*: H^1(\mathfrak{M}_{1,n}) \to H^1(\mathfrak{a})$ is an isomorphism for all n, $\rho^*: H^1(\mathfrak{M}) \to H^1(\mathfrak{a})$ is also an isomorphism.

The proof of the existence of I and a homotopy H is exactly the same as in the simply connected case, see Section 10.■

B. $\pi_1 \otimes \mathbb{Q}$.

Let π be a finitely presented group. We wish to define "$\pi \otimes \mathbb{Q}$". The method is to replace π by a tower of nilpotent groups, $\{N_i(\pi)\}$, and tensor each of them with $\mathbb{Q}$. Thus, $\pi \otimes \mathbb{Q}$ will be a tower of rational nilpotent groups.

Let $\{\Gamma_i(\pi)\}$ be the terms of the lower central series; i.e., $\Gamma_2(\pi) = [\pi,\pi]$; and inductively $\Gamma_{n+1}(\pi) = [\Gamma_n(\pi),\pi]$. Let $N_i(\pi)$ be $\pi/\Gamma_i(\pi)$. Each $N_i(\pi)$ is a nilpotent group of index of nilpotence i; i.e., all i-fold commutators in $N_i(\pi)$ vanish. The $N_i(\pi)$ fit together in a tower:

$$\begin{array}{ccccccc} \cdots \longrightarrow & N_4(\pi) & \longrightarrow & N_3(\pi) & \longrightarrow & N_2(\pi) \\ & \nwarrow & & \uparrow & \nearrow & \\ & & & \pi & & & . \end{array}$$

We have short exact sequences (where $\Gamma_n = \Gamma_n(\pi)$)

$$0 \longrightarrow \Gamma_{n-1}/\Gamma_n \longrightarrow N_n(\pi) \longrightarrow N_{n-1}(\pi) \longrightarrow 1$$

where Γ_{n-1}/Γ_n is in the center of $N_n(\pi)$.

Such extensions correspond to fibrations

$$\begin{array}{ccc} K(\Gamma_{n-1}/\Gamma_n,1) & \longrightarrow & K(N_n(\pi),1) \\ & & \downarrow \\ & & K(N_{n-1}(\pi),1). \end{array}$$

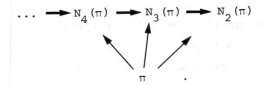

The fact that Γ_{n-1}/Γ_n is central means that in the fibration the fundamental group of the base, $N_{n-1}(\pi)$, acts trivially on the fundamental group of the fiber. (In general, the action is conjugation in $N_n(\pi)$.) Hence, central extensions correspond exactly to principal fibrations with base, total space, and fiber having only fundamental groups. The method for tensoring principal fibrations with $\mathbb{Q}$ leads to a method for tensoring nilpotent groups with $\mathbb{Q}$. If inductively we have:

$$K(N_{n-1}(\pi),1) \longrightarrow K(N_{n-1}(\pi)\otimes \mathbb{Q},1)$$

inducing an isomorphism on rational cohomology, then we define

$$\begin{array}{ccc}
K(\Gamma_{n-1}/\Gamma_n,1) & \longrightarrow & K(\Gamma_{n-1}/\Gamma_n \otimes \mathbb{Q},1) \\
\downarrow & & \downarrow \\
K(N_n(\pi),1) & \longrightarrow & K(N_n(\pi)\otimes \mathbb{Q},1) \\
\downarrow & & \downarrow \\
K(N_{n-1}(\pi),1) & \longrightarrow & K(N_{n-1}(\pi)\otimes \mathbb{Q},1)
\end{array}$$

where the fibration on the right has k-invariant in $H^2(K(N_{n-1}(\pi)\otimes \mathbb{Q},1);\Gamma_{n-1}/\Gamma_n \otimes \mathbb{Q}) \cong H^2(K(N_{n-1}(\pi),1),\Gamma_{n-1}/\Gamma_n)\otimes \mathbb{Q}$ equal to $k \otimes 1_{\mathbb{Q}}$ where k is the k-invariant on the left. Thus we have a ladder of maps between two central extensions:

$$\begin{array}{ccccccccc}
0 & \longrightarrow & \Gamma_{n-1}/\Gamma_n & \longrightarrow & N_n(\pi) & \longrightarrow & N_{n-1}(\pi) & & 1 \\
 & & \downarrow & & \downarrow \gamma_n & & \downarrow \gamma_{n-1} & & \\
0 & \longrightarrow & \Gamma_{n-1}/\Gamma_n \otimes \mathbb{Q} & \longrightarrow & N_n(\pi)\otimes \mathbb{Q} & \longrightarrow & N_{n-1}(\pi)\otimes \mathbb{Q} & \longrightarrow & 1.
\end{array}$$

The extension class for the second one is just the extension class for the first taken with $\mathbb{Q}$-coefficients. Induction on n and a simple comparison theorem of spectral sequences shows that γ_n induces an isomorphism on rational cohomology.

It is a simple argument to show that the elements of finite order in a nilpotent group form a subgroup Tor N. Mal'cev proved that if, N is a nilpotent group, then N/Tor N can be embedded in a uniquely divisible nilpotent group $\mathcal{N}$. (Uniquely divisible means $x^n = a$ has exactly one solution for all $a \in \mathcal{N}$ and $n \in \mathbb{Z}^+$.) If we take $\mathcal{N}$ to be minimal with respect to these properties, then $\mathcal{N}$ is determined up to isomorphism by N. It is $N \otimes \mathbb{Q}$.

Mal'cev[1], however, found $\mathcal{N}$ by associating to N a rational Lie algebra $\mathcal{L}_N$ and then used the Baker-Campbell-Hausdorff formula[2] to define a nilpotent group strucutre, $\mathcal{N}$, on $\mathcal{L}_N$. Note that since $\mathcal{L}_N$ is nilpotent the B-C-H formula becomes a polynomial with rational coefficients and hence defines a group structure on the $\mathbb{Q}$-vector space. This approach has the advantage of proving that the $\mathbb{Q}$-nilpotent groups have $\mathbb{Q}$-nilpotent Lie algebras, and hence are rational algebraic groups. Thus, we can associate to any finitely presented group, π, a tower of rational nilpotent groups, $\{N_n(\pi) \otimes \mathbb{Q}\}$, and tower of rational, nilpotent Lie algebras $\{\mathcal{L}_n(\pi)\}$. These two towers determine each other via the B-C-H formula and its inverse.

Theorem 12.2: Let X be a simplicial complex, and let $\rho_1: \mathfrak{M}_1 \xrightarrow{\rho_1} A(X)$ be a 1-minimal model. Then $\mathfrak{M}_1$ is dual to the tower of Lie algebras $\{\mathcal{L}_n(\pi_1(X))\}$.

Proof: $\mathfrak{M}_1$ is an increasing union of differential algebras: $\mathfrak{M}_{1,1} \subset \mathfrak{M}_{1,2} \subset \dots$. Let V_n be the generators of $\mathfrak{M}_{1,n}$. The differential in $\mathfrak{M}_{1,n}$ is determined by $(d|V_n): V_n \to V_n \wedge V_n$. The rational Lie algebra dual to $\mathfrak{M}_{1,n}$ has underlying vector space $(V_n)^*$. The bracket

(1) See Mal'cev, "On a class of homogeneous spaces", Amer. Math. Soc. Transl. (1) 9 (1962), 276-307.

(2) See G. Baumslag "Lecture notes on nilpotent groups", Regional conference series in Math No. 2.

$$[\ ,\ ]: (V_n)^* \wedge (V_n)^* \longrightarrow V_n^*$$

is dual to $d|V_n$. The Jacobi identity for $[\ ,\ ]$ is dual to the equation $(d^2|V_n) = 0$. We call this Lie algebra structure $\mathcal{L}_n$. The inclusion $i: \mathfrak{M}_{1,n} \subset \mathfrak{M}_{1,n+1}$ gives an inclusion $V_n \subset V_{n+1}$. Dualizing gives

$$0 \longrightarrow K_{n+1} \longrightarrow V_{n+1}^* \xrightarrow{i^*} V_n^* \longrightarrow 0.$$

Since i is a map of D.G.A.'s, i^* is a map of Lie algebras. The fact that $(d|V_{n+1}): V_{n+1} \to V_n \wedge V_n$ dualizes to the fact that K_{n+1} is in the center of the Lie algebra $\mathcal{L}_{n+1}$. Since $\mathcal{L}_0$ is the trivial Lie algebra, it follows by induction that each $\mathcal{L}_n$ is a nilpotent Lie algebra.

This gives the duality between 1-minimal models and towers of nilpotent Lie algebras. It remains to show that this duality carries the 1-minimal model for a simplicial complex to the tower of Lie algebras associated to its fundamental group. This is first proved inductively for towers of principal fibrations of $K(\pi,1)$'s. The proof uses the Hirsch lemma and is the complete analogue of 11.5. Given X, we map it to the following tower:

$$\begin{array}{ccc}
 & & \vdots \\
 & \vdots & \downarrow \\
 & \nearrow & K(N_3(X) \otimes \mathbb{Q}, 1) \\
X & & \downarrow \\
 & \searrow & K(N_2(X) \otimes \mathbb{Q}, 1).
\end{array}$$

The minimal model for the tower is the one dual to the tower of Lie algebras associated with it. On the other hand

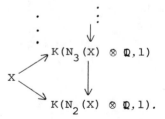

$$H^1(K(N_i(\pi) \otimes \mathbb{Q},1);\mathbb{Q}) = H^1(X:\mathbb{Q})$$

for all i, and

$$\lim\{H^2(K(N_i(\pi) \otimes \mathbb{Q},1);\mathbb{Q})\} \longrightarrow H^2(X;\mathbb{Q})$$

is an injection. Thus, the minimal model for this tower pulls back to the 1-minimal model of X. ■

C. Functorality.

It is when we consider functorality that the basepoint makes its appearance. We define the basepoint of a differential algebra to be a map $A^0 \to k$ (k = ground field). If X is a simplicial complex and $p \in X$ is a basepoint (with rational barycentric coordinates), then it defines $A^0(X) \to \mathbb{Q}$ by evaluation.

If $\mathfrak{A} \to k$ and $\mathfrak{B} \to k$ are differential algebras with basepoints, then a basepoint preserving map $\varphi: \mathfrak{A} \to \mathfrak{B}$ is one that commutes with the maps to k. A homotopy

$$H: \mathfrak{A} \to \mathfrak{B} \otimes (t,dt)$$

is basepoint preserving if

$$\begin{array}{ccc} \mathfrak{A} & \xrightarrow{H} & \mathfrak{B} \otimes (t,dt) \\ \downarrow & & \downarrow \\ k & \hookrightarrow & k \otimes (t,dt) \end{array}$$

commutes. If B^0 equals k, and if H is basepoint preserving, then $H(\alpha) = \Sigma\, b^i \otimes t^i$ for every $\alpha \in \mathfrak{A}^1$. A closer look at the argument given in the proof of theorem 12.1 shows the following:

Theorem 12.3: (a) *If* $\mathfrak{A} \to k$ *is a differential algebra with basepoint, then there is a* 1-*minimal model* $\mathfrak{M}_1$ *(which automatically has a unique basepoint) and a basepoint preserving map*

$$\begin{array}{ccc} \mathfrak{M}_1 & \xrightarrow{\rho} & \mathfrak{A} \\ & \searrow \quad \swarrow & \\ & k & \end{array}$$

inducing an isomorphism on H^1 *and an injection on* H^2.

(b) *Given two such* $\rho: \mathfrak{M}_1 \to \pi$ *and* $\rho': \mathfrak{M}_1' \to \mathfrak{A}$, *then there is an isomorphism* $I: \mathfrak{M}_1 \to \mathfrak{M}_1'$ *(automatically basepoint preserving) and a basepoint preserving homotopy* H *from* $\rho_1' \circ I$ *to* ρ. *The isomorphism* I *is well defined up to basepoint preserving homotopy*.

Completely analogously to 11.2 we have the following:

Theorem 12.4: *If* $f: X \to Y$ *is a base point preserving map, then it induces* $\hat{f}: \mathfrak{M}_{Y,1} \to \mathfrak{M}_{X,1}$ *well defined up to basepoint homotopy*.

At this point a miraculous thing happens.

Lemma 12.5: *If* $H: \mathfrak{M}_1 \to \mathcal{N}_1 \otimes (t,dt)$ *is a base point preserving homotopy between* 1-*minimal algebras, then* $H = H_0 \otimes 1$.

Proof: We prove by induction that $H|\mathfrak{M}_{1,k} = H_0|\mathfrak{M}_{1,k} \otimes 1$. Suppose we know this for $k < n$, and let x be an element of degree $\mathfrak{M}_{1,n}$. Then $H(x) = \Sigma\, n_i \otimes t^i + \omega_i \otimes t^i dt$. Since H is basepoint preserving, all $\omega_i = 0$. Thus $H(x) = \Sigma\, n_i \otimes t^i$ and $H(dx) = \Sigma\, dn_i \otimes t^i + \Sigma\, in_i \otimes t^{i-1}dt$. But $dx \in \mathfrak{M}_{1,n}$ and hence $H(dx) = \alpha \otimes 1$. Thus $in_i = 0$. Consequently $n_i = 0$ for $i > 0$, and hence $H(x) = n_0 \otimes 1$. Since $\mathfrak{M}_n$ is generated by elements in degree 1, it follows that $H|\mathfrak{M} = H_0|\mathfrak{M} \otimes 1$. ∎

Thus we have shown that assigning to basepointed simplicial complexes their basepointed minimal models is a functor from basepointed homotopy theory to 1-minimal models and maps between them. If we dualize to Lie algebras and then exponentiate, the result is the functor that assigns $\pi_1(X) \otimes \mathbb{Q}$ to X and $f_\#: \pi_1(X,p) \otimes \mathbb{Q} \to \pi_1(Y,q) \otimes \mathbb{Q}$ to $f: (X,p) \to (Y,q)$. Thus, from the $\mathbb{Q}$-polynomial forms on X, one has a purely algebraic way to recover the homotopy functor "$\pi_1 \otimes \mathbb{Q}$".

Examples:

1) $S^1 \vee S^1$.

Consider

$$0 \longrightarrow A^* \longrightarrow A^*_{C^\infty}(S^1) \oplus A^*_{C^\infty}(S^1) \xrightarrow{\text{eval at } 0} \mathbb{R} \longrightarrow 0.$$

The differential algebra A^* can be used to calculate the real fundamental group of $S^1 \vee S^1$. There is a map of differential algebras

$$\{H^*(S^1 \vee S^1;\mathbb{R}), d = 0\} \hookrightarrow A^*$$

which induces an isomorphism on cohomology. Hence, to construct the 1-minimal model for A^*, it suffices to construct the 1-minimal model for $H^*(S^1 \vee S^1;\mathbb{R})$. We begin with

$$\Lambda(\{x,y\}) \longrightarrow H^*(S^1 \vee S^1)$$

where x and y are a basis for $H^1(S^1 \vee S^1)$. This map induces an isomorphism on H^1 but it has a kernel in degree 2 generated by $x \wedge y$. The next stage is

$$\rho_2: \Lambda(\{x,y,n\}) \longrightarrow H^*(S^1 \vee S^1)$$

where $dn = x \wedge y$ and $\rho_2(n) = 0$. This map has a kernel in degree 2 generated by $x \wedge n$ and $y \wedge n$. As we continue the construction, the kernels which we encounter in degree 2 keep growing larger and the construction must be repeated ad infinitum. Actually, one is constructing the increasing system of differential algebras dual to the tower of free nilpotent Lie algebras on 2 generators. (Clearly, this is the tower associated to the free group on 2 generators.)

In general, when $H^2(X;\mathbb{Q}) = 0$ the tower of Lie algebras constructed is a tower of free nilpotent Lie algebras.

2) Let N be the nil-manifold obtained by dividing the group of upper triangular real matrices

$$\begin{pmatrix} 1 & x & z \\ 0 & 1 & y \\ 0 & 0 & 1 \end{pmatrix}$$

by the lattice of such integral matrices. This gives a compact 3-manifold whose cohomology ring is the same as that of $S^1 \times S^2 \# S^1 \times S^2$. When we build the 1-minimal model for the forms on $S^1 \times S^2 \# S^1 \times S^2$ we get the infinite process described in Example 1. When we build the 1-minimal model for N, we find that the result is

$$\Lambda(\{x,y\}) \otimes_d \Lambda(n); \quad dn = x \wedge y.$$

Thus $H^2(N)$ is generated by the "Massey Products" $[x \wedge n]$ and $[y \wedge n]$. In fact, the 1-minimal model for N is its minimal model.

XIII. Examples and Computations

A. Spheres and Projective spaces.

Let us consider an odd sphere S^{2n+1}. The first stage in building the minimal model for the forms on S^{2n+1} is to construct an exterior algebra $\Lambda(e)$ on a generator of degree $(2n + 1)$. Clearly, the cohomology of this D.G.A. maps isomorphically to $H^*(S^{2n+1})$ when we send e to a closed form on S^{2n+1} which is not exact. Thus, $\Lambda(e)$ is the minimal model for the forms on S^{2n+1}. By Theorem 11.5, this implies that $\pi_i(S^{2n+1}) \otimes \mathbb{Q}$ is zero for $i \neq 2n+1$ and equal to $\mathbb{Q}$ for $i = 2n+1$.

Let $A^*(S^{2n})$ be the forms on S^{2n}. The first stage in building the minimal model for $A^*(S^{2n})$ is the polynomial algebra on a generator of degree $2n$, $P(x_{2n})$. Clearly, x_{2n}^2 is cohomologous to zero in $A^*(S^{2n})$. Thus, we tensor in an exterior algebra $\Lambda(y_{4n-1})$ with $dy = x^2$. The product $P(x_{2n}) \otimes_d \Lambda(y_{4n-1})$ is the minimal model. It follows that

$$\pi_i(S^{2n}) \otimes \mathbb{Q} \cong \begin{cases} \mathbb{Q} & i = 2n, 4n-1 \\ 0 & i \neq 2n, 4n-1. \end{cases}$$

If $f\colon S^n \to X$ is a simplicial map (with $n > 1$ and X simply connected), the there is induced $\hat{f}\colon \mathfrak{M}_X \to \mathfrak{M}_{S^n}$ well-defined up to homotopy. If we consider $\hat{f}$ in degree n, then it defines a map $I^n(\mathfrak{M}_X)^* \to I^n(\mathfrak{M}_{S^n})^* \cong \mathbb{Q}$. The isomorphism $I^n(\mathfrak{M}_{S^n})^* \xrightarrow{\sim} \mathbb{Q}$ is via integration of the form over the fundamental cycle of S^n. The induced map $I^n(\mathfrak{M}_X)^* \to \mathbb{Q}$ depends only on the homotopy class of $\hat{f}$, and hence only on the homotopy class of f. This defines a map

$$\pi_n(X) \longrightarrow \mathrm{Hom}_{\mathbb{Q}}(I^n(\mathfrak{M}_X), \mathbb{Q}),$$

and hence a map

$$\pi_n(X) \otimes \mathbb{Q} \longrightarrow [I^n(\mathfrak{M}_X)]^*.$$

As we shall see in section XIV, this is exactly the duality between $I^n(\mathfrak{M}_X)$ and $\pi_n(X) \otimes \mathbb{Q}$.

Let us consider the minimal model for the forms on complex projective n-space, $\mathbb{C}P^n$. The first stage of the minimal model is $P(x_2)$. The class $(x_2)^{n+1}$ generates the kernel of the map $P(x_2) \to H^*(\mathbb{C}P^n;\mathbb{Q})$. Thus, the minimal model for the forms on $\mathbb{C}P^n$ is $P(x_2) \otimes_d \Lambda(y_{2n+1})$; $dy = x^{n+1}$. In particular

$$\pi_i(\mathbb{C}P^n) \otimes \mathbb{Q} = \begin{cases} 0 & i \neq 2, 2n+1 \\ \mathbb{Q} & i = 2, 2n+1. \end{cases}$$

One can also deduce this from the calculation of the homotopy groups of S^{2n+1} and the fibration $S^1 \to S^{2n+1} \to \mathbb{C}P^n$.

B. <u>Graded Lie algebras</u>.

Suppose $\mathcal{L} = \oplus_n \mathcal{L}_n$ is a graded vector space (over a field of characteristic 0). Let $[\ ,\ ]: \mathcal{L} \otimes \mathcal{L} \to \mathcal{L}$ be a map which is homogeneous of degree 0. We say that $(\mathcal{L}, [\ ,\ ])$ is a graded Lie algebra if

(1) $[x,y] = (-1)^{(p+1)(q+1)}[y,x]$ for $x \in \mathcal{L}_p$ and $y \in \mathcal{L}_q$ (symmetry).

(2) $[x,[y,z]] = [[x,y],z] + (-1)^{pq}[y,[x,z]]$ for $x \in \mathcal{L}_p$ and $y \in \mathcal{L}_q$ (Jacobi identity).

An ordinary Lie algebra is a graded Lie algebra in which

$\mathscr{L}_n = 0$ for $n \neq 0$.

If X is a simply connected space, then the Whitehead product (see exercise 34) is a bilinear map

$$\pi_p(X) \otimes \pi_q(X) \xrightarrow{[\,,\,]} \pi_{p+q-1}(X).$$

The formulae in exercise 34 show that if we define $\mathscr{L}_n = \pi_{n+1}(X) \otimes \mathbb{Q}$, then the Whitehead product makes $\mathscr{L} = \oplus_{n\geq 1} \mathscr{L}_n$ into a graded Lie algebra.

Another source of examples is D.G.A.'s. Let $\mathfrak{M}$ be a D.G.A which is free as a graded commutative algebra on positive dimensional generators. Let $I(\mathfrak{M}) = \mathfrak{M}^+/\mathfrak{M}^+ \wedge \mathfrak{M}^+$ where $\mathfrak{M}^+$ is the ideal of elements of positive degree. Denote by $(\mathfrak{M}^+)^k$ the kth-power of this ideal; i.e., $(\mathfrak{M}^+)^k = \{\omega | \omega = \Sigma_i\, \alpha_{i1} \wedge \ldots \wedge \alpha_{ik}$ with $\alpha_{ij} \in \mathfrak{M}^+\}$. If $d\colon \mathfrak{M} \to \mathfrak{M}$ is decomposable, then $d\colon (\mathfrak{M}^+)^k \to (\mathfrak{M}^+)^{k+1}$. Hence, it induces a map

$$\begin{array}{ccc} \mathfrak{M}^+/(\mathfrak{M}^+)^2 & \longrightarrow & (\mathfrak{M}^+)^2/(\mathfrak{M}^+)^3 \\ \big\downarrow = & & \big\downarrow = \\ d\colon I(\mathfrak{M}) & \longrightarrow & I(\mathfrak{M}) \wedge I(\mathfrak{M}) \end{array}$$

The fact that $d^2 = 0$ implies that the composition

$$(*)\quad I(\mathfrak{M}) \xrightarrow{d} I(\mathfrak{M}) \wedge I(\mathfrak{M}) \xrightarrow{d\wedge id + (-1)^{\deg} id \wedge d} I(\mathfrak{M}) \wedge I(\mathfrak{M}) \wedge I(\mathfrak{M})$$

is zero. Let $\mathscr{L}_n = [I^{n+1}(\mathfrak{M})]^*$. Dual to d is a map $[\ ,\ ]\colon \mathscr{L} \otimes \mathscr{L} \to \mathscr{L}$ which is homogeneous of degree 0. The fact that d maps into $I(\mathfrak{M}) \wedge I(\mathfrak{M})$ dualizes to the fact that $[\ ,\ ]$ satisfies the symmetry condition to be a graded Lie algebra bracket. The dual to (*) is the Jacobi identity.

There is a connection between these two examples.

If X is a simply connected space, and if $\mathfrak{M}_X$ is its minimal model, then the graded Lie algebras ($\pi_{*+1}(X) \otimes \mathbb{Q}$, Whitehead product) and ($\oplus I^{n+1}(\mathfrak{M}_X)^*, d^*$) are isomorphic under the map

$$\pi_{n+1}(X) \otimes \mathbb{Q} \overset{\simeq}{\to} [I^{n+1}(\mathfrak{M}_X)]^*.$$

As an example of this let us consider $S^2 \vee S^2$. Its minimal model begins with two generators ξ_1 and ξ_2 of degree 2. We then add 3-dimensional generators to kill all 4-dimensional cohomology, and so on. The dual graded Lie algebra to this D.G.A. is the free graded Lie algebra on two 1-dimensional generators ξ_1^* and ξ_2^*.

More generally, $S^{p_1} \vee \ldots \vee S^{p_k}$ has a minimal model whose indecomposables are dual to a free graded Lie algebra on generators of degrees $(p_1-1), \ldots, (p_k-1)$ (assuming each $p_i \geq 2$).

C. The Booromean Rings.

There is a geometric form of Poincaré duality. It allows one to do the calculations involved in building the minimal model with sub-manifolds instead of forms. The basis for the duality is the following. Let $M^k \subset N^n$ be an embedded submanifold. Suppose that M^k and N^n are both closed and oriented. By the tubular neighborhood theorem there is a neighborhood $\nu(M \subset N)$ which is diffeomorphic to a disk bundle over M:

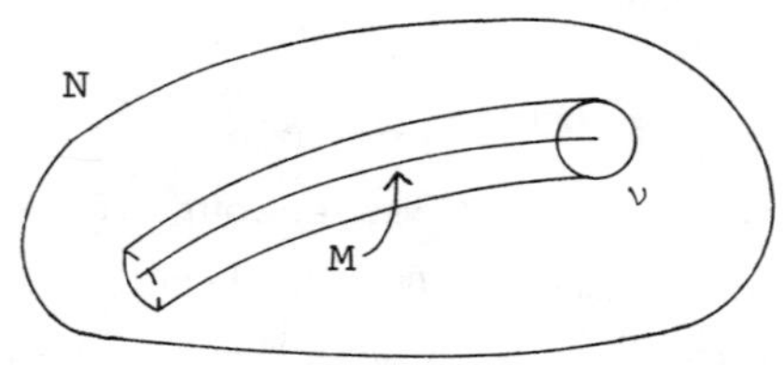

Let D_0^{n-k} be the fiber over a point $m_0 \in M$. Since M and N are oriented, D_0^{n-k} receives an orientation. By the Thom isomorphism theorem (see the exercises) there is a unique class $U_M \in H^{n-k}(\nu(M \subset N), \partial\nu(M \subset N); \mathbb{Z})$ so that $\int_{D_0} U_M = 1$. (Here we assume that M is connected.) There is a C^∞ differential form representing U which vanishes identically near $\partial\nu(M \subset N)$. If we extend by 0 to $N - \nu(M \subset N)$, then we have a closed C^∞-form, $\tilde{U}_M$, on all of N. Its cohomology class is the Poincaré dual of $[M] \in H_k(N)$.

There is a form of this duality for manifolds with boundary. If $(M, \partial M) \subset (N, \partial N)$ with M meeting ∂N normally in ∂M, then the same construction yields a class $\tilde{U}_M \in H^{n-k}(N)$ which is the Lefschetz dual of the class $[M, \partial M] \in H_k(N, \partial N)$.

Under the correspondence $M \to \tilde{U}_M$, transverse intersection of manifolds corresponds to wedge product of forms. Thus, if M_0^k and M_1^ℓ are transverse in N^n with intersection $M_{0,1}^{k+\ell-n}$, and if $\tilde{U}_0$ and $\tilde{U}_1$ are Thom forms for M_0 and M_1 supported in sufficiently small tubes, then $\tilde{U}_0 \wedge \tilde{U}_1$ is a Thom form for $M_{0,1}$. This means that $\tilde{U}_0 \wedge \tilde{U}_1$ is a closed form supported in a tube about $M_{0,1}$ and integrating to 1 over each fiber.

The operation of finding a solution for $d\eta = \tilde{U}_M$, given M and $\tilde{U}_M$, corresponds to finding a sub-manifold of N whose boundary is M. Thus, if $M^k = \partial L^{k+1}$ then there is a form $\tilde{U}_L$ supported in a tube about L, closed outside the tube about M, integrating to 1 over fibers of $\nu(L \subset N)$ which are outside the tube about M, and so that $d\tilde{U}_L = \tilde{U}_M$

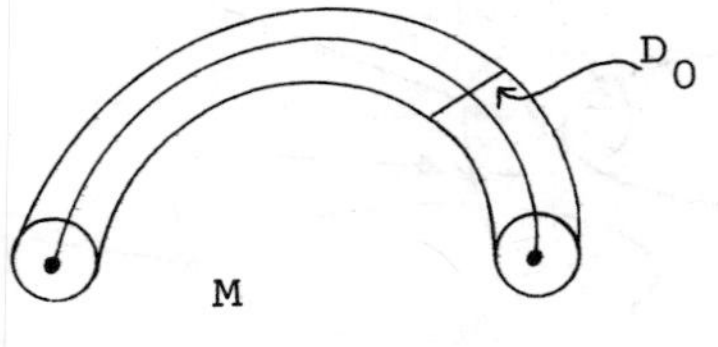

$$\int_{D_0} \tilde{U}_L = 1, \quad d\tilde{U}_L = \tilde{U}_M .$$

We consider now an explicit example of the operation of building the minimal model via sub-manifolds; viz. the ambient space in S^3-(Borromean rings)

$\mathfrak{B}$ = 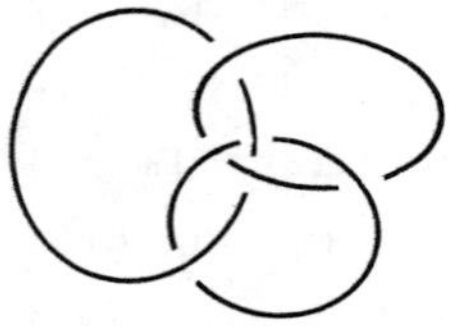 (Borromean rings)

We can think of this as a manifold with boundary by taking out open solid tori around each of the circles, or more simply we work with ordinary cohomology and proper submanifolds.

The first cohomology of $S^3 - \mathfrak{B}$ has rank 3 and is generated by classes which are dual to 2-disks spanning the components. We choose these disks as pictured below:

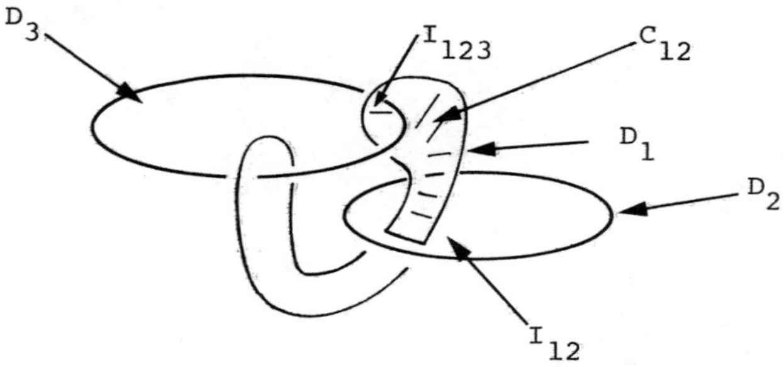

Let $\tilde{U}_1, \tilde{U}_2$, and $\tilde{U}_3$ be the dual Thom classes in $H^1(S^3 - \mathfrak{B})$. The first stage in the minimal model is $\Lambda(\tilde{U}_1, \tilde{U}_2, \tilde{U}_3)$.

The geometric fact that the linking number of any pair is zero in $S^3 - \mathfrak{B}$ means that $\tilde{U}_i \cup \tilde{U}_j = 0$ for all $i \neq j$. Clearly $\tilde{U}_2 \wedge \tilde{U}_3 = 0$ as a form since $D_2 \cap D_3 = \emptyset$. The form $\tilde{U}_1 \wedge \tilde{U}_2$ is the Thom form for the interval I_{12}. To solve the equation $d\xi = \tilde{U}_1 \wedge \tilde{U}_2$ we must find a proper 2-dimensional submanifold whose boundary is I_{12}. We take this to be the part of D_1 cut off by I_{12} which lies above D_2, C_{12}. To form the Massey product $\langle \tilde{U}_1, \tilde{U}_2, \tilde{U}_3 \rangle$ we must take $n_{12} \wedge \tilde{U}_3 + \tilde{U}_1 \wedge n_{23}$ where $dn_{ij} = \tilde{U}_i \wedge \tilde{U}_j$. In this case $n_{23} = 0$ and n_{12} is supported near C_{12}. Thus $n_{12} \wedge \tilde{U}_3 + \tilde{U}_1 \wedge n_{23}$ is represented by the Thom form of $C_{12} \cap D_3$. This intersection is I_{123}. Since I_{123} is an arc with end points on different components of $\mathfrak{B}$, $[I_{123}] \in H_1(S^3 - \mathfrak{B})$ is nonzero. Thus, $\langle \tilde{U}_1, \tilde{U}_2, \tilde{U}_3 \rangle \neq 0$. If we do a similar calculation for $S^3 - \mathfrak{B}'$ where $\mathfrak{B}'$ is three unlinked circles, then all Massey products $\langle \tilde{U}_{i_1}, \tilde{U}_{i_2}, \tilde{U}_{i_3} \rangle$ are trivial. Thus, the third stages of the minimal models for the forms on $S^3 - \mathfrak{B}$ and $S^3 - \mathfrak{B}'$ are different. In particular $\mathfrak{B}$ and $\mathfrak{B}'$ are not isotopic. In fact, since the minimal models differ at the third stage this implies that $\pi_1(S^3 - \mathfrak{B})/\Gamma_5$ and $\pi_1(S^3 - \mathfrak{B}')/\Gamma_5$ are not isomorphic groups. The group $\pi_1(S^3 - \mathfrak{B}')$ is free. The existence of nonzero Massey products in $S^3 - \mathfrak{B}$ means that its fundamental group is not free.

D. Symmetric Spaces and Formality.

A space is said to be formal if the homotopy type of the D.G.A. of forms on the space is the same as the homotopy type of the cohomology ring of the space. Thus, if X is formal and $\mathfrak{M}_X$ is a minimal model for the forms on X, then there is a map $\mathfrak{M}_X \to (H^*(X), d = 0)$ which induces the identity on cohomology.

One can always define a map of cochain complexes

$$(*) \qquad (H^*(X), d = 0) \longrightarrow A^*(X)$$

which induces the identity on cohomology by linearly choosing closed form representatives for each cohomology class. Usually this map will not be multiplicative. If it is possible to choose this map to be multiplicative, then $(H^*(X), d = 0)$ and $A^*(X)$ have the same minimal model. Thus, this gives algebraic topological conditions on a space which must be satisfied if there is to be a multiplicative mapping (*).

If X is a Riemannian manifold, then there is a canonical map $(H^*(X), d = 0) \to A^*(X)$ which assigns to each cohomology class its unique harmonic representative. From the above discussion we see that for X to admit a Riemannian metric in which the wedge product of harmonic forms is harmonic it must be the case that X is formal (over $\mathbb{R}$).

There is one class of Riemannian manifolds in which wedge product of harmonic forms is harmonic. These are the Riemannian locally symmetric spaces.

E. The third homotopy group of a simply connected space.

Let A^* be a D.G.A. with $H^1(A^*) = 0$. The first stage in constructing the minimal model of A^* is the polynomial algebra on $H^2(A^*)$, $\{P[H^2], d = 0\}$. The next stage is created by tensoring in an exterior algebra on the relative 4th-cohomology group $H^4(P[H^2], A^*)$. If X is a simply connected space, then $\pi_3(X) \otimes \mathbb{Q}$ is dual to $H^4(P[H^2], A^*(X))$.

The long exact sequence of the pair $(P[H^2], A^*(X))$ leads to a long exact sequence

$$0 \longrightarrow H^3(X) \xrightarrow{h^*} \mathrm{Hom}(\pi_3(X), \mathbb{Q}) \longrightarrow P_2[H^2(X)] \xrightarrow{U} H^4(X)$$

where h^* is dual to the Hurewicz homomorphism, $P_2[H^2(X)]$ is the vector space of quadratic polynomials in $H^2(X)$, and "U" is the cup product mapping.

If $f: S^3 \to X$ is a continuous mapping, then we can deform it until it becomes simplicial. At this point itinduces a map $f^*: A^*(X) \to A^*(S^3)$, and hence

$$\hat{f}: \mathfrak{M}_X \longrightarrow \mathfrak{M}_{S^3} = \Lambda(e).$$

There is the evaluation map $\mathfrak{M}_{S^3} \longrightarrow \mathbb{Q}$ given by sending $\omega^3 \in \mathfrak{M}_{S^3}$ to $\int_{S^3} \omega^3 \in \mathbb{Q}$. As we said in part B of this section, the map $\pi_3(X) \to [I^3(\mathfrak{M}_X)]^*$ defined by sending $[f]$ to $[\int \circ \hat{f}] \in I^3(\mathfrak{M}_X)^*$ gives the duality between $\pi_3(X)$ and $I^3(\mathfrak{M}_X)$.

Suppose $\omega_1, \ldots, \omega_k \in H^2(X)$ is a basis and $\tilde{\omega}_i$ is a closed 2-form in $A^*(X)$ representing ω_i. Suppose $(\Sigma\, a_{ij}\, \omega_i \wedge \omega_j, \eta) \in P_2[H^2] \oplus A^3(X)$ is a relative cocycle (i.e., $d\eta = \Sigma\, a_{ij}\tilde{\omega}_i \wedge \tilde{\omega}_j$), and that $f: S^3 \to X$. We give a formula for the value of $(\Sigma\, a_{ij}\omega_i \wedge \omega_j, \eta) \in H^4(P_2[H^2], A^*(X)) = I^3(\mathfrak{M}_X)$ on $[f] \in \pi_3(X)$. To do this we must deform

$$\mathfrak{M}_X \xrightarrow{\rho_X} A^*(X) \xrightarrow{f^*} A^*(S^3)$$

by homotopy to factor through $\mathfrak{M}_{S^3} = \Lambda(e)$. Define $H: P[H^2] \to A^*(S^3) \otimes (t,dt)$ as follows: For each $\omega_i \in H^2$, $f^*\tilde{\omega}_i = d\mu_i$ for some 1-form $\mu_i \in A^*(S^3)$. Define $H(\omega_i) = f^*\tilde{\omega}_i \otimes 1 - d(\mu_i \otimes t)$. Let $\hat{f}: (\mathfrak{M}_X)_2 \to \Lambda(e)$ be trivial. At this point we have

$$\begin{array}{ccc} A^*(X) & \xrightarrow{f^*} & A^*(S^3) \\ \uparrow \rho_X & H & \uparrow \rho_{S^3} \\ (\mathfrak{M}_X)_2 & \xrightarrow{0} & \Lambda(e) \end{array}$$

a homotopy commutative diagram.

We wish to extend this to a homotopy commutative diagram in $(\mathfrak{M}_X)_3$. The map $\hat{f}: (\mathfrak{M}_X)_3 \to \Lambda(e)$ must send the class $(\Sigma\, a_{ij}\omega_i \wedge \omega_j, n)$ to some multiple λe. (We assume here that e is chosen so that $\int_{S^3}(\rho_{S^3}(e)) = 1$.) If the given homotopy is to extend over $(\mathfrak{M}_X)_3$, then λ must be equal to

$$\int_{S^3} f^*\eta - \int_0^1 \int_{S^3} H(\Sigma\, a_{ij}\omega_i \wedge \omega_j)$$

$$= \int_{S^3} f^*\eta - \Sigma\, a_{ij}\int_0^1 \int_{S^3}(f^*\tilde{\omega}_i\, 1 - d(\mu_i \otimes t) \wedge (f^*\tilde{\omega}_j \otimes 1) - d(\mu_j \otimes t))$$

$$= \int_{S^3} f^*\eta - \Sigma\, a_{ij}\int_0^1 \int_{S^3}(f^*\tilde{\omega}_i\, 1 - d\mu_i \otimes t + \mu_i \otimes dt)$$

$$\wedge\, (f^*\tilde{\omega}_j\, 1 - d\mu_j \otimes t + \mu_j \otimes dt).$$

The only terms in the second part which will contribute to

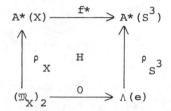

the integral are

$$f^*\widetilde{\omega}_i \otimes 1 \wedge \mu_j \otimes dt - d\mu_i \otimes t \wedge \mu_j \otimes dt + \mu_i \otimes dt \wedge f^*\widetilde{\omega}_j \otimes 1$$
$$- \mu_i \otimes dt \wedge d\mu_j \otimes t$$

$$= f^*\widetilde{\omega}_i \wedge \mu_j \otimes dt - d\mu_i \wedge \mu_j \otimes tdt + \mu_i \wedge f^*\widetilde{\omega}_j\, dt$$
$$- \mu_i \wedge d\mu_j \otimes tdt.$$

Thus,

$$\lambda = \int_{S^3} f^*\eta - \Sigma\, a_{ij} \int_{S^3} f^*\widetilde{\omega}_i \wedge \mu_j + \mu_i \wedge f^*\widetilde{\omega}_j$$
$$+ \Sigma[a_{ij}(\tfrac{1}{2}\int_{S^3} d\mu_i \wedge \mu_j) + \mu_i \wedge d\mu_j]$$

$$= \int_{S^3} f^*\eta - \Sigma[a_{ij}(\int_{S^3} f^*\widetilde{\omega}_i \wedge \mu_j) - \tfrac{1}{2} d\mu_i \wedge \mu_j + \mu_i \wedge f^*\widetilde{\omega}_j$$
$$- \tfrac{1}{2}\mu_i \wedge d\mu_j]$$

$$= \int_{S^3} f^*\eta - \tfrac{1}{2}\Sigma[a_{ij}(\int_{S^3} f^*\widetilde{\omega}_i \wedge \mu_j) + \mu_i \wedge f^*\widetilde{\omega}_j]$$

This formula is a generalization of Whitehead's formula for the Hopf invariant. Whitehead showed that if $f\colon S^3 \to S^2$ is a C^∞ map, then the Hopf invariant $H(f)$ is given by

$$\int_{S^3} f^*\omega \wedge \mu$$

where $\int_{S^2} \omega = 1$ and $d\mu = f^*\omega$. Here, $\pi_3(S^2) \cong \mathbb{Z}$ and the Hopf invariant of an element is its class under this isomorphism.

E. <u>Homotopy theory of certain 4-dimensional complexes</u>.

Let X be a space which is homotopy equivalent to $(\vee_{i=1}^{T} S^2) \cup_{\varphi} e^4$, $\varphi: S^3 \to \vee_{i=1}^{T} S^2$. Any simply connected 4-manifold is so represented. The homotopy type of such a space is determined by $[\varphi] \in \pi_3(\vee_{i=1}^{T} S^2)$. This group is a free abelian group with generators $[x_i, x_j]$ $i \leq j$ where x_i is the identity map of S^2 onto the ith factor in the wedge. (Here, $[\ ,\ ]$ represents the Whitehead product.) There is another description of the element $[\varphi]$ in terms of the cohomology ring of X. This ring is determined by the symmetric pairing

$$H^2(X) \otimes H^2(X) \longrightarrow H^4(X) \xrightarrow[\cong]{\langle\ ,e^4\rangle} \mathbb{Z}.$$

Since $H^2(X)$ has a natural basis dual to the spheres, the cup product mapping is given by a symmetric integral matrix (λ_{ij}). It turns out that $[\varphi] = \Sigma_{i\leq j} \lambda_{ij}[x_i, x_j]$.

This shows that such homotopy types are classified by equivalence classes of symmetric bilinear pairings. The condition that X satisfy Poincaré duality is that the pairing be nonsingular over $\mathbb{Z}$.

It is unknown which nonsingular pairings arise from closed, simply connected 4-manifolds. It is also unknown whether or not, if two such 4-manifolds determine the same pairing, then they are diffeomorphic.

Classifying such complexes up to rational homotopy equivalence is the same as classifying the pairings up to rational equivalence (including a change of scale in the value group). Thus, if X and Y are two simplicial complexes of this form, then $\mathfrak{M}_X$ and $\mathfrak{M}_Y$ are isomorphic if and only if the pairings are rationally equivalent.

The algebras of piecewise C^∞ forms on X and Y have isomorphic minimal models if and only if the pairings for X

and Y are equivalent over the reals (again including an automorphism of the value group). This happens if and only if the pairings are of the same rank, have the same dimensional null space, and have the same signature up to sign.

F. <u>$\mathbb{Q}$-homotopy type of BU_n and U_n</u>, (cf. the exercises).

BU_n is the Grassmannian of n-planes in $\mathbb{C}^\infty$ (thus $BU_n = \lim_{N\to\infty} G(n,N)$). Recall that

$$H^*(BU_n,\mathbb{Z}) \cong \mathbb{Z}[c_1,c_2,\ldots,c_n]$$

where $c_\ell \in H^{2\ell}(BU_n,\mathbb{Z})$ is the ℓth <u>Chern class</u> of the universal vector bundle. This gives a map

$$BU_n \xrightarrow{f} \Pi^n_{\ell=0} K(\mathbb{Z},2\ell)$$

which induces an isomorphism on $\mathbb{Q}$-cohomology. Since $(K(\mathbb{Z},2\ell))_{(0)} = K(\mathbb{Q},2\ell)$ it follows that

$$(BU_n)_{(0)} \xrightarrow{f(0)} \Pi^n_{\ell=0} K(\mathbb{Q},2\ell)$$

is a homotopy equivalence.

<u>Remark</u>: We have thus given a proof of

$$\pi_{2i}(BU_n) \otimes \mathbb{Q} \approx \mathbb{Q}$$

$$\pi_{2i+1}(BU_n) \otimes \mathbb{Q} = 0.$$

This is the <u>unstable version of the Bott periodicity theorem</u>

<u>over</u> $\mathbb{Q}$. In fact, if

$$BU = \lim_{n\to\infty} G(n,2n)$$

then $H^*(BU,\mathbb{Z}) \cong \mathbb{Z}[c_1,c_2,c_3,..,]$. The same argument as above gives that

$$BU_{(0)} \approx \Pi_{\ell=0}^{\infty} K(\mathbb{Q},2\ell).$$

Thus

$$\pi_{2i}(BU) \otimes \mathbb{Q} \approx \mathbb{Q}$$

$$\pi_{2i+1}(BU) \otimes \mathbb{Q} = 0$$

which is the Bott periodicity over $\mathbb{Q}$. The honest Bott periodicity is the result

$$\left.\begin{array}{l} \pi_{2i}(BU) \approx \mathbb{Z} \\ \\ \pi_{2i+1}(BU) = 0 \end{array}\right\}\text{(i)}$$

$$\left.\begin{array}{l} \dfrac{f^*c_n}{(n-1)!} \in H^{2n}(S^{2n},\mathbb{Z}) \\ \\ \text{for any map } S^{2n} \xrightarrow{f} BU \end{array}\right\}\text{(ii)} \quad .$$

The <u>divisibility property</u> (ii) seems to be closely related to deducing B.P./$\mathbb{Z}$ from B.P. over $\mathbb{Q}$.

Let U_n be the unitary group. Recall that

$$H^*(U_n,\mathbb{Z}) \cong \mathbb{Z}\{x_1,x_3,\ldots,x_{2n-1}\}$$

is an exterior algebra with odd-dimensional generators. From this we deduce that the $\mathbb{Q}$-homotopy type

$$(U_n)_{(0)} \approx \Pi_{\ell=1}^{n} S^{2\ell-1}_{(0)} .$$

Thus, over $\mathbb{Q}$, U_n looks like a product of odd spheres. Moreover, we see that

$$\pi_{2i-1}(U_n) \otimes \mathbb{Q} \approx \mathbb{Q} \quad 1 \leq i \leq n$$

$$\pi_{2i}(U_n) \otimes \mathbb{Q} = 0$$

The stable $\mathbb{Z}$ version for $U = \lim_{n\to\infty} U_n$ is

$$\pi_{2i-1}(U) \cong \mathbb{Z}$$

$$\pi_{2i}(U) = 0.$$

This is equivalent to the Bott periodicity above using the exact homotopy sequence of the fibration

$$\begin{array}{ccc} U_n & \longrightarrow & St(n,2n) \\ & & \downarrow \\ & & G(n,2n) \end{array}$$

for n large compared to i ($St(n,2n)$ is the Stiefel manifold--cf. the exercises).

G. <u>Products</u>.

Let A^* and B^* be D.G.A.'s. The tensor product $A^* \otimes B^*$ naturally receives the structure of a D.G.A. If $\mathfrak{M}_A$ and $\mathfrak{M}_B$ are minimal models for A^* and B^*, then $\mathfrak{M}_A \otimes \mathfrak{M}_B$ is a minimal model for $A^* \otimes B^*$. If X and Y are C^∞ manifolds, then there is a map $A^*_{C^\infty}(X) \otimes A^*_{C^\infty}(Y) \to A^*_{C^\infty}(X \times Y)$ which, by the Künneth theorem, induces an isomorphism on cohomology. If X and Y are simplicial complexes and $(X \times Y)'$ is a triangulation of the product cell complex, then there is a map

$$A^*(X) \otimes A^*(Y) \longrightarrow A^*((X \times Y)')$$

which (again by the Künneth theorem) induces an isomorphism on cohomology.

It follows that $\mathfrak{M}_{X\times Y} \cong \mathfrak{M}_X \otimes \mathfrak{M}_Y$. This should be viewed as a generalization of the Künneth theorem. It includes the Künneth theorem (by taking homology). It also includes the rational or real form of the result that $\pi_i(X \times Y) \cong \pi_i(X) \oplus \pi_i(Y)$, since $I(\mathfrak{M}_X \otimes \mathfrak{M}_Y) \cong I(\mathfrak{M}_X) \oplus I(\mathfrak{M}_Y)$.

H. Massey products.

Given a space X and classes

$$\alpha \in H^p(X), \ \beta \in H^q(X), \ \gamma \in H^r(X)$$

that satisfy

$$\alpha \cup \beta = 0 \ \text{ in } \ H^{p+q}(X), \quad \beta \cup \gamma = 0 \ \text{ in } \ H^{q+r}(X),$$

there is defined the Massey triple product

$$\langle\alpha,\beta,\gamma\rangle \in H^{p+q+r-1}(X)/(\alpha \cdot H^{q+r-1}(X) + \gamma \cdot H^{p+q-1}(X)).$$

Using differential forms a,b,c to represent the classes α,β,γ respectively, we write

$$a \wedge b = dg, \quad b \wedge c = dh$$

where g, h are forms of degrees $p + q - 1$, $q + r - 1$. The p + q + r - 1 form

$$k = g \wedge c + (-1)^{p-1} a \wedge h$$

is closed and its cohomology class is well-defined modulo

$$\alpha \cdot H^{q+r-1}(X) + \gamma \cdot H^{p+q-1}(X),$$

as may be directly verified by making different choices in the above recipe. The Massey triple product is then the class in the quotient $H^{p+q+r-1}(X)/(\alpha \cdot H^{q+r-1}(X)+\gamma \cdot H^{p+q-1}(X))$.

I. Massey triple products on compact Kähler manifolds.

We recall the following:

Definition: A Hodge structure of weight m is given by $\mathbb{Q}$-vector space $H_{\mathbb{Q}}$ together with a Hodge decomposition

$$H_{\mathbb{C}} = \oplus_{p+q=m} H^{p,q}$$

$$H^{p,q} = \bar{H}^{q,p}$$

of its complexification $H_{\mathbb{C}} = H_{\mathbb{Q}} \otimes_{\mathbb{Q}} \mathbb{C}$.

Suppose now that M is a compact Kähler manifold (cf. section 7 in chapter 0 of "Principles of Algebraic Geometry", John Wiley (New York), by P. Griffiths and J. Harris for the relevant definitions and notations). If we set $A^{p,q}(M)$ = vector space of complex-valued C^{∞} (p,q) forms on M, and

$$H^{p,q}(M) = \{\varphi \in A^{p,q}(M) : d\varphi = 0\}/dA^{p+q-1}(M) \cap A^{p,q}(M),$$

then it is a basic fact that the subspaces

$$H^{p,q}(M) \subset H^m(M), \qquad p+q=m$$

define a Hodge decomposition on $H^m(M) = H^m(M,\mathbb{Q}) \otimes \mathbb{C}$. Consequently, the cohomology $H^m(M)$ of a compact Kähler manifold has a functorial Hodge structure of weight m (functorial means the obvious thing, to be explained momentarily).

Actually, this statement is a consequence of the following lemma (loc. cit., page 149) about forms on compact Kähler manifolds.

Lemma (principal of two types): On a compact Kähler manifold M, suppose that $\varphi \in A^{p,q}(M)$ is a form such that

$$\varphi = d\eta, \qquad \eta \in A(M).$$

Then we may find forms η' and η'' such that

$$\varphi = d\eta' \quad \text{and} \quad \eta' \in A^{p-1,q}(M)$$

or

$$\varphi = d\eta'' \quad \text{and} \quad \eta'' \in A^{p,q-1}(M).$$

We shall show that the lemma implies that all Massey triple products are zero on a compact Kähler manifold.

For this we remark first that the cup product

$$H^m(M) \otimes H^n(M) \longrightarrow H^{m+n}(M)$$

satisfies the obvious conditions

$$H^{p,q}(M) \otimes H^{r,s}(M) \longrightarrow H^{p+r,q+s}(M)$$

concerning Hodge decompositions.

Thus, in forming the Massey Product $\langle\alpha,\beta,\gamma\rangle$, it suffices to assume that α, β, and γ are themselves homogeneous with respect to the Hodge decomposition; say $\alpha \in H^{t,s}(X)$, $\beta \in H^{i,j}(X)$, and $\gamma \in H^{u,v}(X)$. We shall construct two closed differential forms φ_1 and φ_2 which are cohomologous and which represent $\langle\alpha,\beta,\gamma\rangle$ so that $\varphi_1 \in A^{t+i+u,s+j+v-1}(X)$ and $\varphi_2 \in A^{t+i+u-1,s+j+v}(X)$. In light of the fact that the Hodge decomposition is a direct sum decomposition this will prove that φ_1 and φ_2 are exact, and hence that $\langle\alpha,\beta,\gamma\rangle = 0$.

Choose closed forms $a \in A^{t,s}(X)$, $b \in A^{i,j}(X)$, and $c \in A^{u,v}(X)$ which represent α, β, and γ. We know that $a \wedge b \in A^{t+i,j+s}(X)$ and that it is exact. Choose $y \in A^{t+i,j+s-1}(X)$ and $y' \in A^{t+i-1,j+s}(X)$ so that $dy = dy'$ $= a \wedge b$. The form $y - y'$ is closed. Using the Hodge decomposition in degree $t + s + i + j - 1$, we can vary y and y' by closed forms preserving the bi-graded type so that $y - y'$ becomes exact. Similarly, we choose $z \in A^{i+u,j+v-1}(X)$ and $z' \in A^{i+u-1,j+v}(X)$ so that $dz = dz' = b \wedge c$ and $z - z'$ is exact. The forms $y \wedge c + (-1)^{t+s-1} a \wedge z$ and $y' \wedge c + (-1)^{t+s-1} a \wedge z'$ both represent the Massey product $\langle\alpha,\beta,\gamma\rangle$. The first lies in $A^{s+i+u,t+j+v-1}(X)$ and the second lies in $A^{s+i+u-1,t+j+v}(X)$. Furthermore, they are cohomologous.

Remark: The above is a special case of the theorem proved in "Homotopy theory of Compact Kähler Manifolds" by Deligne, Griffiths, Morgan, and Sullivan, Inv. Math. vol 29, 1974, pp. 245-274) that the homotopy type of a compact Kähler manifold is a formal consequence of the cohomology ring (cf. section D above for the definition of formal). This result, in turn, is generalized to noncompact algebraic varieties in "The Algebraic Topology of Smooth Algebraic Varieties" by Morgan, Publ. IHES vol. 48 (1978).

XIV. Functorality

In section XI we made explicit the duality between minimal D.G.A.'s and rational Postnikov towers. We showed that the minimal model of $A^*(X)$ is dual to the rational Postnikov tower of X. In this section we discuss the functorial properties of this duality.

A. The functorial correspondence.

Let B and C be simplicial complexes (which are simply connected), and let $f: B \to C$ be a continuous mapping. For each vertex v of the triangulation of C, let $\mathring{s}t(v)$ be the open star of v. This is the union of all open simplices of C which have v as a vertex. This gives an open cover of C by $\{\mathring{s}t(v)\}_{v \in \text{vertices}(C)}$. If B' is an subdivision of B so that each simplex of B' lies in $f^{-1}(\mathring{s}t(v))$ for some vertex $v \in C$, then there is a simplicial map $\varphi: B' \to C$ which is homotopic to f. Since the points with rational barycentric coordinates are dense in a simplex, we can always choose B' to be a rational subdivision. The inclusion $I: B' \to B$ induces $A^*(B) \xrightarrow{i^*} A^*(B')$.

Suppose that $\rho_B: \mathfrak{M}_B \to A^*(B)$ and $\rho_C: \mathfrak{M}_C \to A^*(C)$ are given minimal models. The composition $i^* \circ \rho_B: \mathfrak{M}_B \to A^*(B')$ is also a minimal model by Theorem 10.8. There is a map $\hat{f}': \mathfrak{M}_C \to \mathfrak{M}_B$, unique up to homotopy, so that

$$\begin{array}{ccc} A^*(C) & \xrightarrow{(f')^*} & A^*(B') \\ \uparrow{\scriptstyle \rho_C} & & \uparrow{\scriptstyle i^* \circ \rho_B} \\ \mathfrak{M}_C & \xrightarrow{\hat{f}'} & \mathfrak{M}_B \end{array}$$

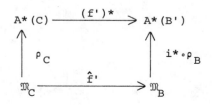

commutes up to homotopy.

Theorem 14.1: *The association* $f \to \hat{f}'$ *defines a functor* $[B,C] \to [\mathfrak{M}_C, \mathfrak{M}_B]$ *which preserves compositions and identities.*

Proof: Let us first show that the homotopy class of $\hat{f}\colon \mathfrak{M}_C \to \mathfrak{M}_B$ depends only on the homotopy class of $f\colon B \to C$. Suppose we have two simplicial approximations $f'\colon B' \to C$ and $f''\colon B'' \to C$ where both B' and B'' are rational subdivisions of B. Form the linear cell complex $B \times I$. There is a rational subdivision of the product cell structure on $B \times I$ which agrees with B' at one end and B'' at the other. If we choose this subdivision fine enough, then there is a simplicial map $H\colon B \times I \to C$ which extends $f' \amalg f''$ on the ends. Thus, there is a commutative diagram:

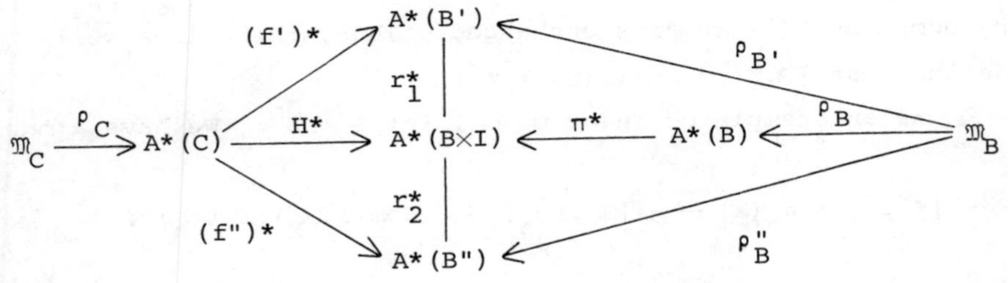

There is $\varphi\colon \mathfrak{M}_C \to \mathfrak{M}_B$ such that $\pi^* \circ \rho_B \circ \varphi$ is homotopic to $H^* \circ \rho_C$. Thus, $\rho_B \circ \varphi$ is homotopic to $(f')^* \circ \rho_C$, and $\rho_{B''} \circ \varphi$ is homotopic to $(f'')^* \circ \rho_C$. Thus, by 10.8, φ is homotopic to $\hat{f}'$ and $\hat{f}''$. This proves that $[B,C] \to [\mathfrak{M}_C, \mathfrak{M}_B]$ is well defined.

Clearly, $[B,B] \to [\mathfrak{M}_B, \mathfrak{M}_B]$ sends the class of the identity to the class of the identity.

Lastly suppose

$$\begin{array}{ccccc} A^*(D) & \xrightarrow{g^*} & A^*(C) & \xrightarrow{f^*} & A^*(B) \\ \uparrow{\scriptstyle \rho_D} & & \uparrow{\scriptstyle \rho_C} & & \uparrow{\scriptstyle \rho_B} \\ \mathfrak{M}_D & \xrightarrow{\hat{g}} & \mathfrak{M}_C & \xrightarrow{\hat{f}} & \mathfrak{M}_B \end{array}$$

is a diagram in which each square homotopy commutes. Since homotopy is an equivalence relation of the maps from a minimal D.G.A.,

$$\rho_B \circ \hat{f} \circ \hat{g} \simeq f^* \circ \rho_C \circ \hat{g} \simeq f^* \circ g^* \circ \rho_D.$$

Thus, $\hat{f} \circ \hat{g}$ is the map associated to $f^* \circ g^*$. This proves that compositions are preserved.■

If $f: \mathfrak{M} \to \mathcal{N}$ is a map between D.G.A.'s then it induces a map on indecomposables $I(f): I(\mathfrak{M}) \to I(\mathcal{N})$. As we vary f by homotopy, $I(f)$ remains unchanged. Thus, there is a well defined map $[\mathfrak{M}, \mathcal{N}] \to \mathrm{Hom}(I(\mathfrak{M}), I(\mathcal{N}))$.

As an example of Theorem 14.1, let $B = S^n$. We have a map

$$[S^n, X] = \pi_n(X) \longrightarrow [\mathfrak{M}_X, \mathfrak{M}_{S^n}] \cong \mathrm{Hom}(I(\mathfrak{M}_X), I(\mathfrak{M}_{S^n})).$$

There is an isomorphism $I^n(\mathfrak{M}_{S^n}) \to \mathbb{Q}$ given by sending $\alpha \in \mathfrak{M}_{S^n}$ to $\int_{S^n} \alpha \in \mathbb{Q}$. Thus, we have a map

$$\pi_n(X) \otimes \mathbb{Q} \xrightarrow{\lambda_X} \mathrm{Hom}(I^n(\mathfrak{M}_X), \mathbb{Q}) = [I^n(\mathfrak{M}_X)]^*.$$

If $f: X \to Y$ is given and $\hat{f}: \mathfrak{M}_Y \to \mathfrak{M}_X$ is an associated map on minimal models, then the following diagram commutes:

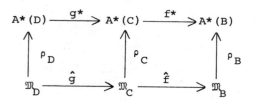

$$\begin{array}{ccc} \pi_n(X) \otimes \mathbb{Q} & \xrightarrow{\lambda_X} & I^n(\mathfrak{M}_X)^* \\ \big\downarrow f_*\otimes \mathrm{Id}_{\mathbb{Q}} & & \big\downarrow I^n(\hat{f})^* \\ \pi_n(Y) \otimes \mathbb{Q} & \xrightarrow{\lambda_Y} & I^n(\mathfrak{M}_Y)^* \end{array} .$$

This follows immediately from the fact that the correspondence in 14.1 is natural and preserves compositions. We shall show later in this section that λ_X is an isomorphism for all simply connected X.

B. <u>Equivalence of the homotopy categories of D.G.A.'s with the rational homotopy category of spaces</u>.

We begin now the proof that if C is a local space, then [B,C] corresponds bijectively with $[\mathfrak{M}_C, \mathfrak{M}_B]$. This is proved by induction on the stages of the Postnikov tower of C. The inductive step requires a detailed analysis of a long exact sequence. It is this analysis that the next 5 propositions develop.

<u>Proposition 14.2</u>: <u>Let</u> $p: E \to B$ <u>be a simplicial map with homotopy theoretic fiber</u> $K(\pi,n)$. <u>Suppose</u> $\pi_1(B) = \{e\}$. <u>Let</u> $\rho_B: \mathfrak{M}_B \to A^*(B)$ <u>be a minimal model. Let</u> $\mathfrak{M}' = \mathfrak{M}_B \otimes_d \Lambda(V)_n$ <u>and suppose there is a map</u> $\rho: \mathfrak{M}' \to A^*(E)$ <u>such that</u>

(1) $\rho|\mathfrak{M}_B = p^* \circ \rho_B$, <u>and</u>

(2) ρ <u>induces an isomorphism on cohomology.</u>

<u>There is a commutative diagram</u>:

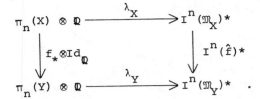

$$\begin{array}{ccc}
[X,B] & \xrightarrow[\text{lifting}]{\text{obst. to}} & H^{n+1}(X;\pi) \\
 & & \downarrow \otimes \mathbb{Q} \\
\Big\downarrow & & H^{n+1}(X;V^*) \\
 & & \downarrow \rho_X^* \\
[\mathfrak{M}_B,\mathfrak{M}_X] & \xrightarrow[\text{extending}]{\text{obst. to}} & H^{n+1}(\mathfrak{M}_X;V^*)
\end{array}$$

Proof: The obstruction to lifting $f: X \to B$ is $f^*(k)$ where $k \in H^{n+1}(B;\pi)$ is the k-invariant of the fibration. The obstruction to extending $\varphi: \mathfrak{M}_B \to \mathfrak{M}_X$ over $\mathfrak{M}'$ is $\varphi^*([d]) \in H^{n+1}(\mathfrak{M}_X;V^*)$. As we saw in 11.5, $[d]$ is the class of $k \otimes 1 \in H^{n+1}(B;\pi) \otimes \mathbb{Q}$ under the identifications $H^*(\mathfrak{M}_B) \xrightarrow{\rho_B^*} H^*(B)$ and $V^* = \mathrm{Hom}(\pi,\mathbb{Q})$. If f and φ correspond, then $f^* = \varphi^*$ under the identifications $H^*(\mathfrak{M}_B) \overset{\rho_B^*}{\to} H^*(B)$ and $H^*(\mathfrak{M}_X) \overset{\rho_X^*}{\to} H^*(X)$. The proposition is immediate from these facts. ■

Proposition 14.3: *Let* $C_{n+1} \overset{p}{\to} C_n$ *be a principal fibration with fiber* $K(\pi,n+1)$. *For any CW-complex* X, *there is an exact sequence:*

$$H^{n+1}(X;\pi) \longrightarrow [X,C_{n+1}] \xrightarrow{p_*} [X,C_n] \xrightarrow{\mathcal{O}} H_{n+2}(X;\pi).$$

This means that $\mathrm{Im}\, p_* = \mathcal{O}^{-1}(0)$, *and that there is an action of* $H^{n+1}(X;\pi)$ *on* $[X,C_{n+1}]$ *so that two maps are in the same orbit if and only if they have the same image in* $[X,C_n]$. *The isotropy group of this action at* $f \in [X,C_{n+1}]$ *is the subgroup* $H_f \subset H^{n+1}(X;\pi)$ *consisting of all elements* α *for which there is a map* $X \times S^1 \overset{\varphi}{\to} C_n$ *with* $\varphi|X \times p = p \circ f$ *and the obstruction to lifting* φ *to* C_{n+1} *is* $\alpha \otimes i \in H^{n+2}(X \times S^1;\pi)$ *where* $i \in H^1(S^1)$ *is the generator*.

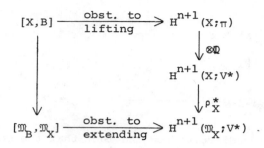

Proof: Since $\mathcal{O}$ is the obstruction to lifting, clearly, $\mathcal{O}^{-1}(0) = \text{Im } p_*$.

Let us define the action of $H^{n+1}(X;\pi)$ on $[X,C_{n+1}]$. Given $\alpha \in H^{n+1}(X;\pi) = H^{n+2}(X \times I, X \times \partial I;\pi)$ and $f\ X \to C_{n+1}$, define a map $\alpha\cdot f: X \to C_{n+1}$ such that $p\circ\alpha\cdot f = p\circ f$ and the obstruction to finding a homotopy of liftings of $p\circ f$ connecting f to $\alpha\cdot f$ is α. One checks easily that this is an action of $H^{n+1}(X;\pi)$ on $[X,C_{n+1}]$. Clearly $p\circ f = p\circ g$ if and only if $[f] = \alpha\cdot[g]$ for some $\alpha \in H^{n+1}(X;\pi)$.

Suppose $\alpha\cdot f = f$ in $[X,C_{n+1}]$. Let $H: X \times I \to C_{n+1}$ be a homotopy form f to αf. Since $p\circ f = p\circ\alpha f$, $p\circ H$ defines a map $\varphi: X \times S^1 \to C_n$. Clearly, $\varphi|X \times \{0\} = p\circ f$. We claim that the obstruction to lifting φ to C_{n+1} is $\alpha \otimes i \in H^{n+2}(X\times S^1;\pi)$. Since we have a lifting $H: X \times I \to C_{n+1}$, the obstruction to lifting φ as a map of $X \times S^1$ lies in $H^{n+1}(X;\pi) \otimes H^1(S^1;\mathbb{Z}) \subset H^{n+2}(X \times S^1;\pi)$. It is easily identified with $\alpha \otimes i$. ■

There is an analogous sequence from Hirsch extensions of D.G.A.'s which we construct now. Let $\mathfrak{M}' = \mathfrak{M} \otimes_d \Lambda(V)_{n+1}$ be a Hirsch extension. Let $\mathfrak{a}$ be a D.G.A.

Proposition 14.4: There is an exact sequence

$$H^{n+1}(\mathfrak{a};V^*) \longrightarrow [\mathfrak{M}',\mathfrak{a}] \xrightarrow{i^*} [\mathfrak{M},\mathfrak{a}] \xrightarrow{\mathcal{O}} H^{n+2}(\mathfrak{a};V^*).$$

This means that $\mathcal{O}^{-1}(0) = \text{Im } i^*$ and that the group $H^{n+1}(\mathfrak{a};V^*)$ acts on $[\mathfrak{M}',\mathfrak{a}]$ so that two elements are in the same orbit if and only if they have the same image in $[\mathfrak{M},\mathfrak{a}]$. The isotropy group for this action at $\varphi \in [\mathfrak{M}',\mathfrak{a}]$ is the subgroup of all $\alpha \in H^{n+1}(\mathfrak{a};V^*)$ for which there is a map $\psi: \mathfrak{M} \to \mathfrak{a} \otimes \Lambda(e)$, with $\pi_{\mathfrak{a}}\circ\psi = \varphi|\mathfrak{M}$, and with the obstruction to extending ψ over $\mathfrak{M}'$ being $\alpha \otimes e \in H^{n+2}(\mathfrak{a} \otimes \Lambda(e);V^*)$.

<u>Proof</u>: Since $\mathcal{O}$ is the obstruction to extending over $\mathfrak{M}'$, $\mathcal{O}^{-1}(0) = \text{Im } i^*$.

Let us define the action of $H^{n+1}(\mathfrak{A};V^*) = \text{Hom}(V,H^{n+1}(\mathfrak{A}))$ on $[\mathfrak{M}',\mathfrak{A}]$. Given $\alpha\colon V \to H^{n+1}(\mathfrak{A})$ and $\varphi\colon \mathfrak{M}' \to \mathfrak{A}$, we define $\alpha\cdot\varphi$. On $\mathfrak{M} \subset \mathfrak{M}'$, $\alpha\cdot\varphi = \varphi$. Choose a lifting $\tilde{\alpha}\colon V \to \{\text{closed forms of } \mathfrak{A}^{n+1}\}$ for α. Define $\tilde{\alpha}\cdot\varphi\colon V \to \mathfrak{A}$ by $\alpha\cdot\varphi(v) = \varphi(v)+\tilde{\alpha}(v)$. One sees easily that this defines a map of D.G.A.'s. The homotopy class of $\tilde{\alpha}\cdot\varphi$ depends only on the homotopy class of φ and the cohomology class of $\tilde{\alpha}$. Let us show that φ_1, φ_2 have the same image under i^* if and only if there is $\alpha \in H^{n+1}(\mathfrak{A},V^*)$ so that $\alpha\varphi_1 = \varphi_2$. The "if" direction is immediate from the definition of the action. Suppose, conversely, that we have maps φ_1 and φ_2 so that $\varphi_1|\mathfrak{M}$ is homotopic to $\varphi_2|\mathfrak{M}$. Consider the diagram:

$$\begin{array}{ccc}
\mathfrak{M} & \xrightarrow{\varphi_1|} & \mathfrak{A} \\
\big\downarrow & & \| \\
\mathfrak{M}' = \mathfrak{M} \otimes_d \Lambda(V)_{n+1} & \xrightarrow{\varphi_2} & \mathfrak{A}
\end{array} .$$

It is homotopy commutative. Since $H^*(\mathfrak{A},\mathfrak{A}) = 0$, there is no obstruction to finding $\varphi_2'\colon \mathfrak{M}' \to \mathfrak{A}$ which is homotopic to φ_2 and such that $\varphi_2'|\mathfrak{M} = \varphi_1|\mathfrak{M}$.

Thus, we may assume that $\varphi_1|\mathfrak{M} = \varphi_2|\mathfrak{M}$. For each $v \in V$, consider $\varphi_1(v) - \varphi_2(v) \in \mathfrak{A}^{n+1}$. Since $dv \in \mathfrak{M}$, $d\varphi_1(v) = \varphi_1(dv) = \varphi_2(dv) = d\varphi_2(v)$. Hence $\varphi_1(v) - \varphi_2(v)$ is closed. If $\alpha\colon V \to H^{n+1}(\mathfrak{A})$ is the resulting homomorphism then one sees easily that $\alpha\cdot\varphi_2 = \varphi_1$ in $[\mathfrak{M}',\mathfrak{A}]$.

Lastly, we need to understand the isotopy groups of this action. Let $\tilde{\alpha}\colon V \to \{\text{closed forms of } \mathfrak{A}^{n+1}\}$ be a representative for α. Let $\varphi\colon \mathfrak{M}' \to \mathfrak{A}$. If $\tilde{\alpha}.\varphi = \varphi$ in $[\mathfrak{M}',\mathfrak{A}]$, then there is a homotopy $H\colon \mathfrak{M}' \to \mathfrak{A} \otimes (t,dt)$ from φ to $\tilde{\alpha}\cdot\varphi$. If we restrict H to $\mathfrak{M}$, it becomes a homotopy from $\varphi|\mathfrak{M}$ to $\varphi|\mathfrak{M}$.

Let $K \subset (t,dt)$ be all forms $\Sigma\, a_i t^i + b_i t^i dt$ such that $\Sigma_{i\geq 1}\, a_i = 0$. This is a sub-D.G.A. The algebra $\mathfrak{A} \otimes K \subset \mathfrak{A} \otimes (t,dt)$ is the kernel of $r_1 - r_0\colon \mathfrak{A} \otimes (t,dt) \to \mathfrak{A}$ where r_i is restriction at $t = i$.

<u>Lemma 14.5</u>: $\Lambda(dt) \subset K$ <u>induces an isomorphism on cohomology</u>.

This is a straight-forward computation.

As a corollary we have that $j\colon \mathfrak{A} \otimes \Lambda(e) \to \mathfrak{A} \otimes K$, defined by $e \to dt$, induces an isomorphism on cohomology.

Consider the diagram

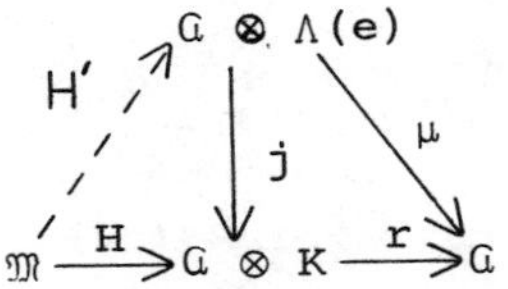

where $r\colon \mathfrak{A} \otimes K \to \mathfrak{A}$ is restriction at $t = 0$, and μ is defined by setting $e = 0$. The obstructions to lifting H up to homotopy vanish. Thus, there is $H'\colon \mathfrak{M} \to \mathfrak{A} \otimes \Lambda(e)$ so that $\mu \circ H' = \varphi$, and so that $j \circ H'$ is homotopic to H.

There is no obstruction to extending H' to a homotopy $J'\colon \mathfrak{M}' \to \mathfrak{A} \otimes (t,dt)$ such that

(1) $J'|\mathfrak{M} = H'$ and

(2) J' is a homotopy from φ to $\tilde{\alpha}\cdot\varphi$.

We claim that the obstruction to extending H' to a map $\mathfrak{M}' \to \mathfrak{A} \otimes \Lambda(e)$ is exactly the homomorphism $v \to [\tilde{\alpha}(v) \otimes e] \in H^{n+2}(\mathfrak{A} \otimes \Lambda(e))$. To see this note that the obstruction to extending H' sends $v \in V$ to the class $H'(dv) \in H^{n+2}(\mathfrak{A} \otimes \Lambda(e))$. Since $J'\colon \mathfrak{M}' \to \mathfrak{A} \otimes (t,dt)$ is an extension of $j \circ H'|\mathfrak{M}$, $H'(dv) = J'(dv) = dJ'(v)$. Thus $j \circ H'(dv)$ is of the form $a_v \otimes 1 + b_v \otimes dt$. Suppose

$J'(v) = \Sigma\, a_{v,i} \otimes t^i + b_{v,i} \otimes t^i dt$. We see that

$$\Sigma\, da_{v,i} \otimes t^i + (-1)^{\deg(a_{v,i})} i a_{v,i} \otimes t^{i-1}dt + db_{v,i} \otimes t^i dt$$

$$= a_v \otimes 1 + b_v \otimes dt.$$

This implies that $a_v = da_{v,0}$, and hence that a_v is exact. Since

$$d\int_{t=0}^{t=1} H(v) + \int_{t=0}^{t=1} H(dv) = H\Big|_{t=1}(v) - H\Big|_{t=0}(v) = \widetilde{\alpha}(v)$$

we have

$$(-1)^{\deg(b_v)} b_v = \widetilde{\alpha}(v) - d\Big(\int_{t=0}^{t=1} H(v)\Big).$$

Thus in cohomology the class of $H'(dv)$ is $(-1)^{\deg(b_v)}(\widetilde{\alpha}(v) \otimes e)$. This proves that, up to sign, the obstruction to extending H' is the homomorphism $v \to [\widetilde{\alpha}(v) \otimes e] \in H^{n+2}(\mathcal{G} \otimes \Lambda(e))$. ■

Let $p\colon C_{n+1} \to C_n$ be a simplicial model for a principal fibration with fiber $K(\pi, n+1)$. Let $\rho_n\colon \mathfrak{M}_n \to A^*(C_n)$ be a minimal model. Let $\mathfrak{M}_{n+1} = \mathfrak{M}_n \otimes_d \Lambda(V)_{n+1}$ and let $\rho_{n+1}\colon \mathfrak{M}_{n+1} \to A^*(C_{n+1})$ be a map extending $p^* \circ \rho_n$ and inducing an isomorphism on cohomology. Let X be a simplicial complex, and let $\rho_X\colon \mathfrak{M}_X \to A^*(X)$ be a minimal model. There is a commutative diagram of exact sequences:

$$\begin{array}{ccccccc} H^{n+1}(X;\pi) & \longrightarrow & [X, C_{n+1}] & \longrightarrow & [X, C_n] & \longrightarrow & H^{n+2}(X;\pi) \\ \downarrow \rho_X^* & & \downarrow & & \downarrow & & \downarrow \rho_X \\ H^{n+1}(\mathfrak{M}_X; V^*) & \longrightarrow & [\mathfrak{M}_{n+1}, \mathfrak{M}_X] & \longrightarrow & [\mathfrak{M}_n, \mathfrak{M}_X] & \longrightarrow & H^{n+2}(\mathfrak{M}_X; V^*). \end{array} \tag{14.6}$$

The only point that is left to check is the compatibility of the actions of $H^{n+1}(X;\pi)$ on $[X,C_{n+1}]$ and of $H^{n+1}(\mathfrak{M}_X;V^*)$ on $[\mathfrak{M}_{n+1},\mathfrak{M}_X]$.

Suppose $\alpha \in H^{n+1}(X;\pi)$ and $f\colon X \to C_{n+1}$ are given. The maps

$$\mathfrak{M}_{n+1} \xrightarrow{\rho_{n+1}} A^*(C_{n+1}) \xrightarrow{f^*} A^*(X)$$

and

$$\mathfrak{M}_{n+1} \xrightarrow{\rho_{n+1}} A^*(C_{n+1}) \xrightarrow{(\alpha f)^*} A^*(X)$$

are homotopic on $\mathfrak{M}_n$. The obstruction to extending the homotopy over $\mathfrak{M}_{n+1}$ is exactly $\alpha \in H^{n+1}(X;\pi \otimes \mathbb{Q})$. Thus, if φ_f and $\varphi_{\alpha f}$ are maps $\mathfrak{M}_{n+1} \to \mathfrak{M}_X$ representing f and αf, then they are homotopic on $\mathfrak{M}_n$, and the obstruction to extending the homotopy over $\mathfrak{M}_{n+1}$ is α. Hence, $\varphi_{\alpha f} = \alpha\cdot\varphi_f$ in $[\mathfrak{M}_{n+1},\mathfrak{M}_X]$.

<u>Theorem 14.7</u>: <u>Let</u> C <u>be a local space. The functor</u> $[X,C] \to [\mathfrak{M}_C,\mathfrak{M}_X]$ <u>is a bijection.</u>

<u>Proof</u>: The proof is by induction on the Postnikov tower of C. We show that $[X,C_n] \to [\mathfrak{M}_C(n),\mathfrak{M}_X]$ is a bijection for all n, where C_n is the nth stage in the Postnikov tower for C and $\mathfrak{M}_C(n)$ is the subalgebra of $\mathfrak{M}_C$ generated in degrees $\leq n$. The inductive step follows by the 5-lemma from the exact sequence 14.6 together with the fact that the identification $H^{n+1}(X;\pi) \xrightarrow{\rho_X^*} H^{n+1}(\mathfrak{M}_X;V^*)$ induces group isomorphisms on the corresponding isotropy groups.■

<u>Corollary 14.8</u>: <u>The functorial mapping</u> $\lambda_X\colon \pi_n(X) \otimes \mathbb{Q} \to I^n(\mathfrak{M}_X)^*$ <u>is an isomorphism for all simply connected spaces</u>.

<u>Proof</u>: We show that λ_X is onto. Let $\varphi\colon I^n(\mathfrak{M}_X) \to \mathbb{Q}$ be given. There is a map of D.G.A.'s $\hat{\varphi}\colon \mathfrak{M}_X \to \mathfrak{M}_{S^n}$ which realizes $\hat{\varphi}$. It

is clear how to define $\hat{\varphi}$ in degrees $\leq n$. In the higher degrees there are no obstructions to extending $\hat{\varphi}$ since $H^*(\mathfrak{M}_{S^n}) = 0$ for $* > n$. Once we have $\hat{\varphi}$, theorem 14.7 tells us there is $\varphi\colon S^n \to X_{(0)}$ which realizes $\hat{\varphi}$. Thus, $\pi_n(X_{(0)}) \xrightarrow{\lambda_{X_{(0)}}} I^n(\mathfrak{M}_X)^*$ is onto. But $\pi_n(X_{(0)}) = \pi_n(X) \otimes \mathbb{Q}$, and by naturality $\lambda_{X_{(0)}} = \lambda_X \otimes Id_{\mathbb{Q}}$.

To see that λ_X is 1-1 suppose $\lambda_X(f) = 0$. This implies that $\hat{f}\colon \mathfrak{M}_X \to \mathfrak{M}_{S^n}$ induces the zero map on indecomposables in degree n. Elementary obstruction theory implies that $\hat{f}$ itself is homotopic to zero. Hence, by 14.7, $S^n \xrightarrow{f} X \subset X_{(0)}$ is homotopically trivial. This means that $[f] \in \pi_n(X)$ is trivial in $\pi_n(X_{(0)}) = \pi_n(X) \otimes \mathbb{Q}$.■

<u>Corollary 14.9</u>: <u>Let</u> X <u>and</u> Y <u>be simply connected spaces and</u> $f\colon X \to Y$ <u>a continuous map. From</u> $\mathfrak{M}_X$ <u>and</u> $\mathfrak{M}_Y$ <u>we can construct spaces</u> $X_{(0)}$ <u>and</u> $Y_{(0)}$ <u>which are localizations for</u> X <u>and</u> Y. <u>From</u> $\hat{f} \in [\mathfrak{M}_Y, \mathfrak{M}_X]$ <u>we can construct a map</u> $f_{(0)}\colon X_{(0)} \to Y_{(0)}$ <u>which is a representative, up to homotopy, of the localization of</u> f.

Appendix A1

In this appendix we shall prove the Hirsch lemma.

Let X be a topological space. Let $\mathrm{Cube}_*(X)$ be the singular cubical chain complex on X. Let $Q_*(X)$ be the quotient chain complex of nondegenerate cubes; i.e., $Q_n(X) = \mathrm{Cube}_n(X)/D_n(X)$ where $D_n(X)$ is the free abelian group generated by maps $\varphi: I^n \to X$ which factor through $p: I^n \to I^{n-1}$, $p(t_1,\ldots,t_n) = (t_1,\ldots,t_{n-1})$. It is easy to see that there are isomorphisms:

$$H_*(\mathrm{Cube}_*(X)) \cong H_*(\mathrm{Sing}_*(X))$$

and

$$H_*(\mathrm{Cube}_*(X)) \cong H_*(Q_*(X)).$$

Thus $Q_*(X)$ is a chain complex which calculates the singular homology.

Let $C^*(X)$ denote the D.G.A. of $\mathbb{Q}$-polynomial forms on $Q_*(X)$. We view the elements as collections of forms $\{\omega_\sigma\}_{\sigma\in \mathrm{Cube}_*(X)}$ such that:

(1) if τ is a face of σ then $\omega_\sigma|\tau = \omega_\tau$ and

(2) if σ is a degenerate cube then $\omega_\sigma = 0$.

Similar to the p.l. de Rham theorem we have the following:

<u>Theorem</u>: <u>Integration defines a map</u>

$$\int: C^*(X) \longrightarrow Q^*(X;\mathbb{Q})$$

<u>which induces an isomorphism on cohomology</u>.

<u>Corollary</u>: <u>If</u> X <u>is a simplicial complex, then there is an isomorphism, well defined up to homotopy, between the minimal model of</u> C*(X) <u>and that of</u> A*(X).

<u>Proof of Corollary</u>: Define

$$A^*(\mathrm{Sing}_*(X)) = \{\omega_\sigma \in A^*(\sigma) \mid \tau < \sigma \Rightarrow \omega_\tau = \omega_\sigma|\tau\}$$

where σ and τ run over the singular simplices of X and $A^*(\sigma)$ means the $\mathbb{Q}$-polynomial forms on the domain of σ.

$$\widetilde{C}^*(X) = \{\omega_c \in A^*(c) \mid c' < c \Rightarrow \omega_{c'} = \omega_c|c'\}$$

where c and c' run over the singular cubes of X and A*(c) means the $\mathbb{Q}$-polynomial forms on the domain of c.

$$(p\ell\,\widetilde{C})^*(X) = \{\omega_c \in B^*(c) \mid c' < c \Rightarrow \omega_{c'} = \omega_c|c'\}$$

where c and c' run over the singular cubes of X and B*(c) means that D.G.A. of all forms which are piecewise $\mathbb{Q}$-polynomial forms on some rational subdivision of the domain of c.

The barycentric subdivision map

singular cube of $X \longmapsto$ sum of singular simplices of X

induces a map of DGA's $A^*(\mathrm{Sing}_*(X)) \to (p\ell\, \tilde{C})^*(X)$. The inclustion of the simplices of X as singular simplices induces a projection of DGA's

$$A^*(\mathrm{Sing}_*(X)) \longrightarrow A^*(X) \longrightarrow 0.$$

We have the inclusion of DGA's

$$\tilde{C}^*(X) \hookrightarrow (p\ell\, \tilde{C})^*(X).$$

Lastly, there is the projection of DGA's

$$\tilde{C}^*(X) \longrightarrow C^*(X).$$

All these maps are seen to induce isomorphisms on cohomology. Hence, $A^*(X)$ and $C^*(X)$ have minimal models which are identified by an isomorphism well-defined up to homotopy.∎

Let $p: E \to B$ be a fibration. Let $\sigma: I^n \to E$ be a cube. The composite $p \circ \sigma: I^n \to B$ can be factored uniquely as $\pi_k: I^n \to I^k$, $(t_1,\ldots,t_n) \to (t_1,\ldots,t_k)$, followed by a non-degenerate map $b_\sigma: I^k \to B$. Let A^*_σ denote the $\mathbb{Q}$-polynomial forms on the domain of σ. The above factoring induces a decomposition $I^n = I^k \times I^{n-k}$. There is an induced decomposition $A^*_\sigma \cong A^*_{b_\sigma} \otimes A^*_{f_\sigma}$ where $A^*_{b_\sigma} = A^*(I^k \times \{0\})$ and $A^*_{f_\sigma} = A^*(\{0\} \times I^{n-k})$. We define a decreasing filtration $F(A^*_\sigma)$ by:

$$F^p(A^*_\sigma) = \Sigma_{i \geq p}\, A^i_{b_\sigma} \otimes A^*_{f_\sigma}.$$

Clearly,

$$d: F^p(A^*_\sigma) \longrightarrow F^p(A^*_\sigma).$$

Furthermore, the restriction mapping preserves this filtration (though it does not preserve the tensor product splitting).

We define $F(C^*(E))$ by letting $F^p(C^*(E))$ be all compatible collections

$$\{\omega_\sigma \in F^p(A^*_\sigma)\}.$$

Since $d\colon F^p(C^*(E)) \to F^p(C^*(E))$, we have a spectral sequence. Since $F^p(C^*(E)) \otimes F^q(C^*(E)) \overset{\wedge}{\to} F^{p+q}(C^*(E))$, this spectral sequence has a multiplicative structure.

Let us calculate E_0 and d_0 for each $\sigma \in Q(E)$. First note tha $F^p(A^*_\sigma)/F^{p+1}(A^*_\sigma) \cong A^p_{b_\sigma} \otimes A^{*-p}_{f_\sigma}$. Clearly, under this identification $d_0\colon F^p/F^{p+1} \to F^p/F^{p+1}$ becomes $(-1)^p \otimes d_{f_\sigma}$.

<u>Lemma</u>: <u>Let</u> F <u>be the fiber</u> $p^{-1}(b_0)$. <u>Suppose</u> $\pi_1(B,b_0)$ <u>acts trivially in</u> $H^*(F)$. <u>There is an identification</u> $E_1^{p,q} \cong C^p(B) \otimes H^q(F)$.

<u>Proof</u>: Let $\alpha \in E_0^{p,q}$. For each cube $\sigma\colon I^n \to E$, the form α_σ is in $A^p_{b_\sigma} \otimes A^q_{f_\sigma}$. If α is a d_0-cocycle, then α_σ is actually in $A^p_{b_\sigma} \otimes Z^q_{f_\sigma}$ where $Z^q_{f_\sigma} \subset A^q_{f_\sigma}$ is the subspace of closed forms. If $x \in H_q(F)$ is a given class, then we define $\langle\alpha, x\rangle \in C^p(B)$ as follows: For each nongenerate simplex $\tau \in Q(B)$, let F_τ be the fiber over $\tau(0,0,\ldots,0)$. Homotopy lifting yields

$$\begin{array}{ccc} I^n \times F_\tau & \longrightarrow & E \\ \Big\downarrow & p_1 & \Big\downarrow p \\ I^n & \longrightarrow & B \end{array} .$$

Since $\pi_1(B,b_0)$ acts trivially on F, we can identify F with F_τ uniquely up to homotopy, by choosing a path from b_0 to $\tau(0,\ldots,0)$. This gives

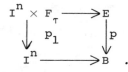

$$\begin{array}{ccc} I^n \times F & \xrightarrow{\tilde{\tau}} & E \\ \downarrow p_1 & & \downarrow p \\ I^n & \xrightarrow{\tau} & B \end{array} .$$

The element $x \in H_q(V)$ is represented by a nondegenerate cubical cycle $\Sigma\, \lambda_i \mu_i$, $\mu_i\colon I^q \to F$. Each of these μ_i yields $\sigma_i\colon I^n \times I^q \to E$ so that $b_{\sigma_i} = \tau$ and $f_{\sigma_i} = \mu_i$. Write $\alpha_{\sigma_i} = \Sigma_j\, \beta_{i,j} \otimes f_{i,j}$ where $\beta_{i,j} \in A^p_\tau$ and $f_{i,j} \in A^q_{\mu_i}$ is a closed form. The value of α on x is

$$(-1)^p\, \Sigma_{i,j}\, \lambda_i \beta_{ij} \cdot \int_{\mu_i(I^q)} f_{i,j}.$$

<u>Claim:</u> This gives a well defined map

$$Z_0^{p,q} \longrightarrow C^p(B) \otimes H^q(F)$$

which vanishes on $\operatorname{Im} d_0$.

<u>Proof</u>: A homomorphism $H_q(F) \to C^p(B)$ is, by duality, identified with an element in $C^p(B) \otimes H^q(F)$. The choices we made in defining $\langle\alpha,x\rangle$ were

(1) to define $I^n \times F \to E$ covering τ and inducing a homotopy equivalence on each fiber, and

(2) to choose a cycle representative for x.

If we change these choices, the resulting homotopy class in the fibers $p^{-1}(p)$, $p \in \tau$, is unchanged. Thus, the value of $\langle\alpha,x\rangle$ is independent of these choices. Clearly if $\alpha \in \operatorname{Im} d_0$ then $\langle\alpha,x\rangle = 0$ since the forms f_{ij} are exact.

Thus we have a well defined map

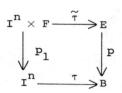

$$E_1^{p,q} \longrightarrow C^p(B) \otimes H^q(F).$$

It is easy to see that this map is an isomorphism, and that d_1 is $d_B \otimes 1$ so that $E_2^{p,q} = H^p(B;H^q(F))$. ■

Now suppose $E \overset{p}{\to} B$ is a fibration with fiber $K(\pi,n)$. Then we claim that <u>there is a map of</u> D.G.A.'s

$$C^*(B) \otimes_d \Lambda(\pi^*) \longrightarrow C^*(E)$$

<u>inducing an isomorphism on cohomology</u>. Together with the above corollary, this will complete the proof of 11.2. First notice that p induces a map $Q(E) \to Q(B)$, and hence a map $p^*: C^*(B) \to C^*(E)$. Since the fiber of p is $K(\pi,n)$, the relative cohomology of $(C^*(B),C^*(E))$ is zero in degrees $\leq n$ and $H^{n+1}(C^*(B),C^*(E)) \cong \mathrm{Hom}(\pi,\mathbb{Q}) = \pi^*$. Define a map $\pi^* \to C^{n+1}(B) \oplus C^n(E)$ whose image lies in the space of relative $(n+1)$ cocycles and which represents the identity $\pi^* \to H^{n+1}(C^*(B),C^*(E))$.

This induces $C^*(B) \otimes_d \Lambda(\pi^*) \to C^*(E)$ extending p^* on $C^*(B)$. Define a filtration on $C^*(B) \otimes_d \Lambda(\pi^*)$ by

$$F^p(C^*(B) \otimes \Lambda(\pi^*)) = \oplus_{i\geq p} C^i(B) \otimes_d \Lambda(\pi^*).$$

The map $C^*(B) \otimes_d \Lambda(\pi^*) \to C^*(E)$ preserves the filtrations. It thus induces maps of the spectral sequences. Clearly $E_2^{p,q}(C^*(B) \otimes_d \Lambda(\pi^*)) = H^p(B;\Lambda(\pi^*)^q) = H^p(B;H^q(F))$. The map on the E_2 level is multiplicative. It is the identity on $E_2^{p,0} = H^p(B)$ and is an isomorphism on $E_2^{0,n} = \pi^*$. By one of the exercises this implies that the map at $E_2^{p,q}$ is an isomorphism, and hence that $C^*(B) \otimes_d \Lambda(\pi^*) \to C^*(E)$ induces an isomorphism on cohomology. ■

<u>Corollary</u>: <u>If</u> $\mathfrak{M}_B$ <u>is the minimal model for</u> B, <u>then there is</u>

a map $\mathfrak{M}_B \otimes_d \Lambda(\pi^*) \to C^*(E)$ inducing an isomorphism on cohomology.

Corollary: If $p: E \to B$ is a simplicial model for a fibration with fiber $K(\pi,n)$, and if $\mathfrak{M}_B$ is the minimal model for $A^*(B)$, then there is a map $\mathfrak{M}_B \otimes_d \Lambda(\pi^*) \to A^*(E)$ inducing an isomorphism on cohomology.

Proof: This follows immediately from the previous corollary and the fact that $A^*(X)$ and $C^*(X)$ have the same minimal model.■

Appendix A2. Functorial comparison of the C^∞- and $\mathbb{Q}$-polynomial minimal models.

If M is a C^∞ manifold with a C^∞ triangulation, then we have a diagram:

$$A^*(M) \otimes_{\mathbb{Q}} \mathbb{R} \hookrightarrow \tilde{A}^*(M) \longleftarrow A^*_{C^\infty}(M)$$

where both inclusions induce isomorphisms on cohomology. Let the three minimal models be denoted $\mathfrak{M}_{\mathbb{Q}}(M)$, $\mathfrak{M}(M)$, and $\mathfrak{M}_{C^\infty}(M)$. We have isomorpisms, well defined up to homotopy

$$\mathfrak{M}_{\mathbb{Q}}(M) \otimes_{\mathbb{Q}} \mathbb{R} \xrightarrow{\cong} \tilde{\mathfrak{M}}(M) \xleftarrow{\cong} \mathfrak{M}_{C^\infty}(M).$$

It is the purpose of this appendix to show that these identifications are natural. Thus, if $f\colon M \to N$ is a C^∞-map, and if $g\colon M \to N$ is a simplicial approximation (in some C^∞ triangulations), then the induced maps $\hat{f}\colon \mathfrak{M}_{C^\infty}(N) \to \mathfrak{M}_{C^\infty}(M)$ and $\hat{g}\colon \mathfrak{M}_{\mathbb{Q}}(N) \to \mathfrak{M}_{\mathbb{Q}}(M)$ are compatible with the identifications in the sense that, under the identifications, $\hat{g} \otimes 1_{\mathbb{R}}$ is homotopic to $\hat{f}$.

To prove this we need the following geometric theorem.

<u>Theorem</u>: <u>Let</u> $f\colon M \to N$ <u>be a</u> C^∞ <u>mapping. There is a</u> C^∞ <u>triangulation of</u> N, K_N, <u>and a</u> C^∞ <u>triangulation of</u> M, K_M, <u>so that if</u> $\sigma \in K_M$, <u>then</u> $f(\sigma)$ <u>is contained in a closed simplex of</u> K_N. <u>Furthermore, there is a homotopy</u> $H\colon M \times I \to N$ <u>from</u> f <u>to a simplicial map so that</u> $H|\sigma \times I$ <u>is contained in a closed simplex of</u> K_N <u>and</u> $H|\sigma \times I$ <u>is</u> C^∞.

<u>Proof</u>: Take any C^∞ triangulation of N. By Sard's theorem

if we deform it slightly, we can ensure that f is transverse to the triangulation. For each closed simplex τ in K_N, $f^{-1}(\tau)$ is a manifold with "corners". In fact, $X_\tau = f^{-1}(\tau)$ has boundary equal to $\bigcup_{\substack{\tau'<\tau \\ \dim \tau'=\dim\tau-1}} X_{\tau'}$ and the $\{X_{\tau'}\}$ meet like the faces $\{\tau'\}$ meet in τ. We triangulate M in a C^∞ fashion so that each X_τ becomes a subcomplex. Let K_M be the triangulation. If $\sigma \in K_M$, then $\sigma \subset X_\tau$ for some τ, and hence $f(\sigma) \subset \tau$.

Define a function, φ, from the vertices of K_M to those of K_N so that any time v (a vertex of K_M) is contained in X_τ, $\varphi(v)$ is a vertex of τ. We claim that φ determines a linear map $g: K_M \to K_N$. To show this, we must show that if $\{v_0,\ldots,v_k\}$ are vertices of a simplex σ of K_M, then $\{\varphi(v_0),\ldots,\varphi(v_k)\}$ are vertices of some simplex of K_N. But any simplex σ in K_M is contained in some X_τ. Thus $\varphi(v_i)$ is a vertex of τ for $i = 0,\ldots,k$.

Now we construct a homotopy $H: M \times I \to N$ so that

(1) $H_0 = \mathrm{Id}$ and $H_1 = g$

(2) $H(X_\tau \times I) \subset \tau$ and

(3) $H|(\sigma \times I)$ is C^∞ for any simplex σ of K_M.

Suppose inductively that we have defined H, as required, on $(\bigcup_{\dim \tau'<s} X_{\tau'}) \times I$. Let τ be a simplex of K_N of dimension s. Since $\partial X_\tau = \bigcup_{\tau'<\tau} X_{\tau'}$, we have already defined $H: \partial X_\tau \times I \to \tau$. We will extend this as required, over all of X_τ. Doing this for the various X_τ for which $\dim \tau = s$ will complete the induction.

Let us consider now X_τ. Suppose inductively that we have extended the map H to a map, still called H, on $(\partial X_\tau \cup (X_\tau)^{(r-1)}) \times I$ as required. Let σ^r be an r-simplex in X_τ which is not in ∂X_τ. We have $f \cup H \cup g: \partial(\sigma^r \times I) \to \tau$, a map which is C^∞ on every (closed) face. Suppose that such a map can be extended to a C^∞ map $H_\sigma\colon \sigma^r \times I \to \tau$. Choosing such extensions, one for each σ^r in X_τ will construct H as required on $\partial X_\tau \cup (X_\tau)^r$, and hence, by induction on all of X_τ.

Lemma: Given C^∞-functions $f_1,\ldots,f_n$ on the coordinate hyperplanes $\{x_1 = 0\},\ldots,\{x_n = 0\}$ so that f_i and f_j agree on $\{x_1 = 0\} \cap \{x_j = 0\}$, there is a C^∞-function on $\mathbb{R}^n$ whose restriction to $\{x_i = 0\}$ is f_i for all $i = 1,\ldots,n$.

Proof: If I denotes the multi-index $(i_0,\ldots,i_r)$ with $i_0 < i_1 < \ldots < i_r$, then let $|I|$ be r, and let $\pi^I\colon \mathbb{R}^n \to \mathbb{R}^n$ be the projection of $\mathbb{R}^n$ onto the subspace of dimension $n-|I|-1$ made up of the coordinate directions different from $i_0,\ldots,i_r$. The formula for the function f is

$$f(x) = -\Sigma_{r=0}^{n} \Sigma_{I,|I|=r} (-1)^{|I|} f_{i_0}(\pi^I(x)) \qquad \blacksquare$$

Corollary: Given $f \cup H \cup g\colon \partial(\sigma^r \times I) \to \tau$, a map which is C^∞ on each closed face, there is a C^∞ extension $H_\sigma\colon \sigma \times I \to \tau$.

Proof: By the previous lemma (and the compactness of $\partial(\sigma \times I)$) we can find finitely many open sets $\{U_1,\ldots,U_s\}$ of $\sigma \times I$ which cover $\partial(\sigma \times I)$ so that each U_i has a C^∞ mapping $H_i\colon U_i \to \tau$ extending the given map on $U_i \cap (\partial(\sigma \times I))$. Let $U_0 = \mathrm{int}(\sigma \times I)$ and let $H_0\colon U_0 \to \hat{\tau}$ (where $\hat{\tau}$ is the barycenter of τ). Take a C^∞ partition of unity subordinate to $\{U_0,\ldots,U_s\}$, $\{\lambda_i\}$, and define

$$\bar{H}\colon \sigma \times I \longrightarrow \tau$$

by $\bar{H}(p,t) = \Sigma_{i=0}^{n} \lambda_i(p,t) H_i(p,t)$. (This gives an element of τ since $\Sigma_{i=0}^{n} \lambda_i(p,t) = 1$ and $0 \leq \lambda_i(p,t)$.)

Clearly, $\bar{H}$ is a C^∞ function. If $(p,t) \in \partial(\sigma \times I)$ then for every i for which $\lambda_i(p,t) \neq 0$, $H_i(p,t) = f \cup H \cup g(p,t)$. Hence, $\bar{H}(p,t) = f \cup H \cup g(p,t)$. ■

This completes the proof of the theorem. ■

Since $H: \sigma \times I \to \tau$ is C^∞, it induces

$$H^*: \widetilde{A}^*(N) \longrightarrow \widetilde{A}^*(M \times I)$$

provided that we use the triangulation K_N to define $\widetilde{A}^*(N)$ and the cellulation $K_M \times I$ to define $\widetilde{A}^*(M \times I)$. This gives commutative diagrams

$$\begin{array}{ccc}
A^*_{C^\infty}(N) & \xrightarrow{f^*} & A^*_{C^\infty}(M) \\
\big\downarrow & & \big\downarrow \\
 & & \widetilde{A}^*(M) \\
 & & \big\uparrow{\scriptstyle t=0} \\
\widetilde{A}^*(N) & \xrightarrow{H^*} & \widetilde{A}^*(M \times I)
\end{array}$$

and

$$\begin{array}{ccc}
A^*(N) \otimes_{\mathbb{Q}} \mathbb{R} & \xrightarrow{g^* \otimes 1_{\mathbb{R}}} & A^*(M) \otimes_{\mathbb{Q}} \mathbb{R} \\
\big\downarrow & & \big\downarrow \\
 & & \widetilde{A}^*(M) \\
 & & \big\uparrow{\scriptstyle t=1} \\
\widetilde{A}^*(N) & \xrightarrow{H^*} & \widetilde{A}^*(M \times I)
\end{array} .$$

Since the restriction mappings induce isomorphisms in cohomology, they give identifications, well defined up to homotopy:

$$\mathfrak{M}_{\mathbb{Q}}(M) \otimes_{\mathbb{Q}} \mathbb{R} \xrightarrow{\cong} \widetilde{\mathfrak{M}}(M \times I) \xleftarrow{\cong} \mathfrak{M}_{C^\infty}(M).$$

Clearly under this identification, and the identifications

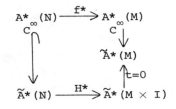

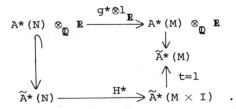

$$\mathfrak{M}_{\mathbb{Q}}(N) \otimes_{\mathbb{Q}} \mathbb{R} \xrightarrow{\cong} \widetilde{\mathfrak{M}}(N) \longleftarrow \mathfrak{M}_{C^\infty}(N) ,$$

f^*, H^*, and g^* induce homotopic mappings. This proves the result stated at the beginning of this appendix.

Exercises

1. If X, Y are finite C.W. complexes, then prove that $X \times Y$ is also. Show that the k-skeleton

$$(X \times Y)^{(k)} = \bigcup_{i+j=k} X^{(i)} \times Y^{(j)}.$$

From this, deduce that the cellular chain groups

$$\tilde{C}_*(X \times Y) \cong \tilde{C}_*(X) \otimes \tilde{C}_*(Y)$$

with boundary

$$\partial_{X\times Y} = \partial_X \otimes 1 \pm 1 \otimes \partial_Y.$$

Using this prove the Künneth theorem.
Why is the Künneth theorem hard for singular or simplicial homology?

2. Let (X,A) be a CW pair where A is a finite subcomplex. Show that

$$X/A = \{X \text{ with } A \text{ collasped to the base point}\}$$

is again a C.W. complex. Is the same thing true if A is infinite? Show that

$$H_*(X,A) \cong \tilde{H}_*(X/A) \quad \text{(use excision)}$$

$$\pi_n(X,A) \neq \pi_n(X,A) \text{ in general}$$

(Hint: Take X = 2-disc and A = S^1 its boundary. Then $X/A \cong S^2$ and so $\pi_3(X/A) \neq 0$. What about $\pi_3(X,A)$?)

3. (a) Show that $\pi_0(X) = \{\text{set of arc components of } X\}$.

(b) Let ΩX be the loop space of X. Show that

$$\pi_0(\Omega X) \cong \pi_1(X).$$

(c) In general, prove that

$$\pi_{n-1}(\Omega X) \cong \pi_n(X),$$

and deduce that if by induction we define $\Omega^n X = \Omega(\Omega^{n+1} X)$, then

$$\pi_0(\Omega^n X) \cong \pi_n(X).$$

4. Let X be a CW complex and $f\colon S^n \to X^{(n)}$ a map. Let $Y = X \cup_f e^{n+1}$ be the CW complex obtained by attaching the n+1-cell by f. Show that the homotopy type of Y depends only on the homotopy class

$$[f] \in \pi_n(X^{(n)}).$$

(<u>Note</u>: We are not considering $[f]$ in $\pi_n(X)$. Why not?)

5. Given spaces X, Y define the <u>wedge</u>

$$X \vee Y = (X \times y_0) \cup (x_0 \times Y) \subset X \times Y.$$

Compute the cohomology <u>ring</u> of $\mathbb{C}P^{n-1} \vee S^{2n}$. Is this the same as the cohomology ring of $\mathbb{C}P^n$?

6. Using the preceeding exercise, show that

$$\pi_{2n-1}(\mathbb{C}P^{n-1}) \neq 0.$$

7. Using the exact homotopy sequence, show that

$$\pi_n(X \times Y) \cong \pi_n(X) \oplus \pi_n(Y).$$

8. Let X be a space, $\tilde{X} \xrightarrow{f} X$ its universal covering. Using the homotopy lifting property, prove that

$$\pi_n(\tilde{X}) \xrightarrow{f_*} \pi_n(X)$$

is an isomorphism for $n \geq 2$. Using this, let

$X =$ 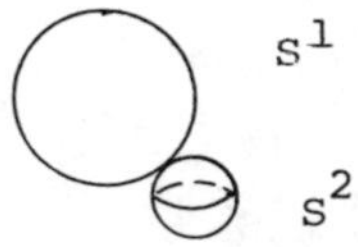

(= 2 sphere with wire attached).

Show that $\pi_2(X) \cong \mathbb{Z} \oplus \mathbb{Z} \oplus \ldots$ *infinite* number of times. Thus, the homotopy groups of a finite complex may be *infinitely generated*. [Is π_1 (finite complex) finitely generated?]

9. Consider the space $X = \{x = 0\} \cup \{y = \sin \frac{1}{x}\}$ in the (x,y)-plane. Show that $\pi_0(X)$ has two elements, but X is connected as a topological space. Let $Y = \{a\} \cup \{b\}$ be a

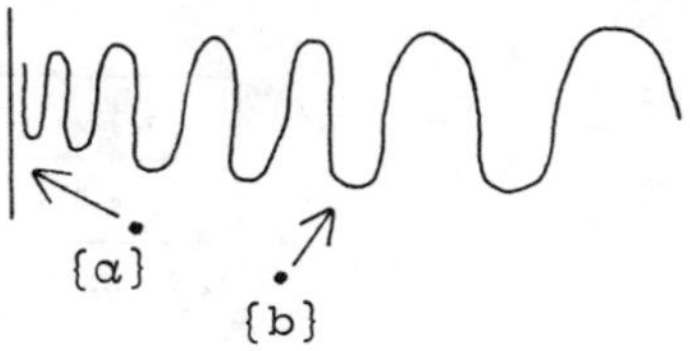

space with two distinct points and $f: Y \to X$ the above map in the picture. Show that f_* is an isomorphism on homotopy groups, but f^{-1} does not exist in the homotopy category. Thus the Whitehead theorem is false for non-CW complexes.

10. a) Show that the inclusion

$$S^n \vee S^n \hookrightarrow S^n \times S^n$$

induces an isomorphism

$$\pi_n(S^n \vee S^n) \cong \pi_n(S^n \times S^n) \quad (n > 1).$$

(Hint: Using exercise 7, the map is onto. If we have

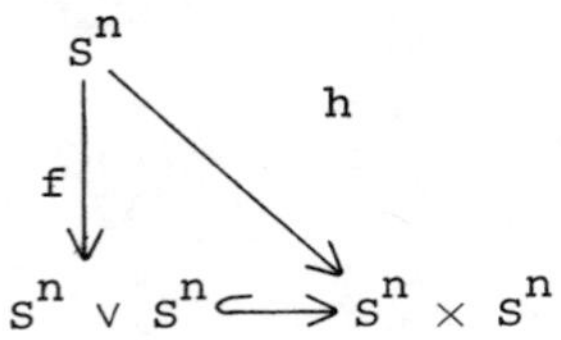

such that h is homotopic to a constant, then since $\dim(S^n \times S^n) = 2n > n+1$, we may assume that the homotopy misses a point $p \in S^n \times S^n$ (why? cf. the simplicial approximation theorem). Thus you must prove that

<u>Lemma</u>: $S^n \times S^n - p$ <u>retracts onto</u> $S^n \vee S^n$.

For this write $S^n = I^n/\partial I^n$ and use a picture like:

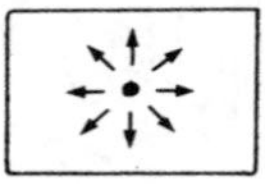

$I^n/\partial I^n$

$I^n/\partial I^n$

b) What is $\pi_1(S^1 \vee S^1)$ (picture = ∞)?

11. a) Fill in the details of the proof of Theorem 3.2.

b) Why does the proof fail for $n = 1$?

12. Given $A = B \cup_f e^{n+1}$ show that

$$\pi_i(B) \cong \pi_i(A) \qquad i < n$$

$$\pi_n(B) \longrightarrow \pi_n(A) \longrightarrow 0$$

(Hint: Show that it is sufficient to check that any map $X^{(n)} \stackrel{f}{\to} A$ from an n-complex to A is homotopic to a map missing a point in e^{n+1}. Now why is this true? See exercise 10 above.)

13. Using exercise 4 and the simplicial approximation theorem, show that any (finite) CW complex has the same homotopy type as a simplicial complex.

(Hint: Given X, suppose that we have an n-dimensional simplicial complex $Y^{(n)}$ and a homotopy equivalence $X \underset{g}{\stackrel{\sim}{\to}} Y^{(n)}$. Let $Y^{(n)} \cup_f e^{n+1}$ be an n+1 cell attached to $Y^{(n)}$. Make a simplicial approximation h to $g \circ f$ and check that $Y^{(n)} \cup_h e^{n+1}$ is a simplicial complex and there is a homotopy equivalence

$$X \cup_f e^{n+1} \sim Y^{(n)} \cup_h e^{n+1}.$$

Now continue adding more cells.)

14. A space X is of <u>category two</u> if

$$X = X_+ \cup X_- \quad (X_+, X_- \text{ open in } X)$$

$$X_+, X_- \text{ are contractible to a point.}$$

Show that, for any space Y, the <u>suspension</u> SY is of category two. In particular, any sphere S^n is of category two (this is obvious).

15. Let X be of category two and $Y = X_+ \cap X_-$.

Show that the sets

$$\mathrm{Vect}^r(X)$$

and

$$[Y, GL(r,\mathbb{C})]$$

are in 1-1 correspondence ($\mathrm{Vect}^r(X)$ = vector bundles over X with fiber $\mathbb{C}^r$).

(Hint: Setting $E_+ = E|X_+$ and $E_- = E|X_-$, both $E_+ \to X_+$ and $E_- \to X_-$ are trivial (why)? Now compare the trivializations over $X_+ \cap X_- = Y$.)

16. An r-frame in $\mathbb{C}^N$ is given by r orthonormal vectors $(e_1,\ldots,e_r)$. The set of r-frames in $\mathbb{C}^N$ is the Stiefel manifold $S(r,N)$. Show that

$$\pi_i(S(r,N)) = 0 \qquad i < 2N-2r+1$$

$$\pi_{2N-2r+1}(S(r,N)) \cong \mathbb{Z}.$$

(Hint: Observe first that $S(1,N) = S^{2N-1}$ is the unit sphere $\mathbb{C}^N$, so the result is correct for $r = 1$. Now use induction on r and the exact homotopy sequence of the fibering

$$\begin{array}{ccc} S^{2N-2r+1} & \longrightarrow & S(r,N) \\ & & \Big\downarrow \pi \\ & & S(r-1,N) \end{array}$$

where π is given by $\pi(e_1,\ldots,e_r) = (e_1,\ldots,e_{r-1})$.)

17. Show that the set of all r-frames in $\mathbb{C}^r$ is the unitary group $U(r)$ (the r-frames in $\mathbb{C}^r$ are just the orthonormal bases).

Deduce the fibration

$$\begin{array}{ccc} U(r) & \longrightarrow & S(r,N) \\ & & \Big\downarrow \pi \\ & & G(r,N) \end{array}$$

where $G(r,N)$ is the Grassmann manifold of r-planes in $\mathbb{C}^N$ and

$$\pi(e_1,\ldots,e_r) = e_1 \wedge \ldots \wedge e_r$$
$$= \text{r-plane spanned by } e_1,\ldots,e_r.$$

Using this fibration show that

$$\pi_i(G(r,N)) \cong \pi_{i-1}(U(r)) \quad \text{for} \quad i < 2N-2r+1.$$

18. Use the inclusion $U(n) \hookrightarrow U(n+1)$ given in matrices by

$$A \longrightarrow \begin{pmatrix} A & 0 \\ 0 & 1 \end{pmatrix}$$

$$A = n \times n \text{ unitary matrix.}$$

Deduce that there is a fibering

$$\begin{array}{ccc} U(n) & \longrightarrow & U(n+1) \\ & & \Big\downarrow \pi \\ & & S^{2n+1} \end{array}$$

and conclude that

$$\pi_i(U(n)) \cong \pi_i(U(n+1)) \quad \text{for} \quad i < 2n.$$

Thus the homotopy of the unitary group is <u>stable</u> in the sense that

$$\pi_i(U(n)) \cong \pi_i(U(n+m)) \quad (i < n \text{ and } m \geq 0).$$

Letting $U = \lim_{n\to\infty} U(n)$, it is a famous theorem of Bott (<u>periodicity theorem</u>) that

$$\begin{cases} \pi_{2i+1}(U) \cong \mathbb{Z} \\ \pi_{2i}(U) = 0 \end{cases}$$

In section XII this theorem is proved in the rational homotopy category.

(<u>Note</u>: Let $G(n,2n)$ be the Grassmannian of n-planes in $\mathbb{C}^{2n}$. Using exercise 17 and Bott periodicity, we have

$$\pi_{2i}(G(n,2n)) \cong \mathbb{Z} \qquad (n \gg i)$$

$$\pi_{2i-1}(G(n,2n)) = 0$$

The map $\pi_{2i}(G(n,2n)) \xrightarrow{\rho} \mathbb{Z}$ is given as follows: A homotopy element in $\pi_{2i}(G(n,2n))$ is given by $S^{2i} \to G(n,2n)$. Pulling back the universal bundle ζ on $G(n,2n)$ gives a vector bundle $\mathbb{C}^n \to E \to S^{2i}$. Then

$$\rho(f) = c_i(E)/(i-1)! \in \mathbb{Z}$$

where $c_j(E) \in H^{2j}(S^{2i})$ is the jth Chern class of E and we have identified $H^{2i}(S^{2i}) \cong \mathbb{Z}$. This divisibility property of the Chern classes of a vector bundle is extremely subtle, and is the sort of thing not covered by rational homotopy theory.

19. Complete the proof of Corollary 3.6 as follows: Given a CW complex X with $\pi_1(X) = 0$ and $H_i(X) = 0$ for $i > n$ and $H_n(X)$ free, look at the (n-1)-skeleton $X^{(n-1)}$. Using that

$$H_i(X^{(n-1)}) \cong H_i(X) \quad i \leq n-2$$

$$H_{n-1}(X^{(n-1)}) \longrightarrow H_{n-1}(X) \longrightarrow 0$$

deduce the exact sequence

$$0 \longrightarrow H_n(X) \longrightarrow H_n(X,X^{(n-1)}) \longrightarrow H_{n-1}(X^{(n-1)}) \xrightarrow{i} H_{n-1}(X) \longrightarrow 0.$$

Now $H_{n-1}(X^{(n-1)})$ is free (why?), and so is $H_n(X)$ by assumption. Thus $H_n(X,X^{(n-1)})$ is free (why?). Show finally that $\pi_i(X,X^{(n-1)}) = 0$ for $i \leq n-1$ and $\pi_n(X,X^{(n-1)}) \cong H_n(X,X^{(n-1)})$, so that we may attach n cells $\{e^n_\alpha\}$ to have $Y = X^{(n-1)} \cup_{f_\alpha} \{e^n_\alpha\}$ together with a map $Y \to X$ inducing an $\approx$ on homology.

20. Here is an example of CW complexes X, Y and a map

$$X \xrightarrow{f} Y$$

such that f_* is isomorphism on $H_*(X) \overset{\cong}{\to} H_*(Y)$ but f is not a homotopy equivalence. Consider

$$Y_1 = S^1 \vee S^2$$

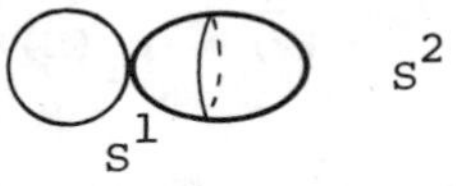

The universal covering $\widetilde{Y}_1$ has a picture

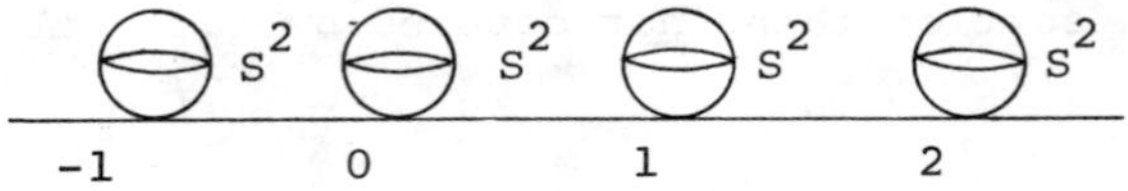

where $\mathbb{R}$ is the real line (= universal covering of S^1) and 2-spheres are attached at each integer point. Let $T\colon \widetilde{Y}_1 \to \widetilde{Y}_1$ be translation by 1 on $\mathbb{R}$ viewed as an automorphism of $\widetilde{Y}_1$. Thus $Y_1 = \widetilde{Y}_1/\{T\}$ where T generates $\pi_1(Y_1)$ as a group of covering transformations on $\widetilde{Y}_1$.

Attach a 3-cell to $\widetilde{Y}_1$ by a map

$$S^2 \xrightarrow{\tilde{f}} \widetilde{Y}_1$$

and let the attaching map to Y_1 be $\pi \circ \tilde{f} = f$

$$\begin{array}{ccc} S^2 & \xrightarrow{\tilde{f}} & \widetilde{Y}_1 \\ & \searrow^{f} & \downarrow \pi \\ & & Y_1 \end{array} \quad .$$

To give $\tilde{f}$, we map

S^2 to $S^2_{\{T=0\}}$ with degree +2 and to $S^2_{\{T=1\}}$ with degree -1

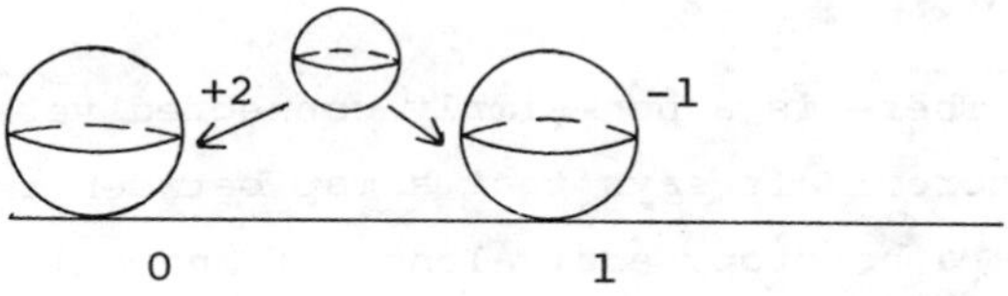

Let $Y = Y_1 \cup_f e^3$. Then

$$H_i(Y) = 0 \quad i \neq 1$$

$$H_1(Y) \cong \mathbb{Z} \qquad \text{(why?)},$$

and so the inclusion map

$$X = S^1 \overset{i}{\hookrightarrow} Y$$

induces $\approx$ on H_*, and even on π_1. But $\pi_2(X) = 0$ and we will check that $\pi_2(Y) \neq 0$ so that i is not a homotopy equivalence.

To do this, observe that the homotopy groups $\pi_i(\tilde{Y}_1)$ are $\pi_1(Y_1)$-modules (why?) (the homology $H_*(\tilde{Y}_1)$ is also a $\pi_1(Y_1)$ module). Letting g be the generator of $\pi_2(\tilde{Y}_1)$ corresponding to the S^2 at 0, we have as sets

$$\pi_2(\tilde{Y}_1) = \{T^n g^m\}_{n,m\in\mathbb{Z}} \qquad \text{(why?)}.$$

Thus, as groups

$$\pi_2(\tilde{Y}_1) \cong \mathbb{Z}[T] \qquad \text{(why?)}$$

Now

$$\pi_2(Y) \cong \pi_2(\tilde{Y}) \cong H_2(\tilde{Y}) \qquad \text{(why?)}.$$

Finally,

$$H_2(\tilde{Y}) \cong \mathbb{Z}[T]/2T - 1 \neq 0$$

(since $\partial e^3 = 2S_0^2 - S_1^2$) (why?).

Remark: (i) There is a non-simply connected version of the Hurewicz theorem. It says that a map between two CW complexes, $f: X \to Y$, is a homotopy equivalence if and only if:

1) $f_*: \pi_1(X) \to \pi_1(Y)$ is an isomorphism, and

2) $f_*: H_*(X;\mathbb{Z}[\pi_1(X)]) \to H_*(Y;\mathbb{Z}[\pi_1(Y)])$ is an isomorphism.

The point is that $H_*(X;\mathbb{Z}[\pi_1(X)])$ is identified with the homology of the universal cover $\tilde{X}$ of X. Thus condition 2) implies that $\tilde{f}: \tilde{X} \to \tilde{Y}$ induces an isomorphism in homology and hence an homotopy (since $\pi_1(\tilde{X}) = \pi_1(\tilde{Y}) = 0$).

(ii) The dodecahedron group G is a perfect group (i.e., $G = [G,G]$) acting freely on S^3, and so $X = S^3/G$ is a 3-manifold with $H_i(X) = 0$ $(i \neq 0,3)$, $H_0(X) \cong H_3(X) \cong \mathbb{Z}$. Thus X is a homology sphere, and the map

$$X \xrightarrow{f} S^3$$

obtained by collapsing the 2-skeleton of X to a point induces $\approx$'s on homology. This example, due to Poincaré, gives another example of exercise 20, and shows that the homology does not suffice to determine the homotopy type of even a 3-manifold.

21. Problem on algebraic Euler characteristics. Recall that a finitely generated abelian group G has a rank $\rho(G) = \dim(G \otimes_{\mathbb{Z}} \mathbb{R})$. Suppose that (C_n,∂) is a chain complex with C_n finitely generated and $C_n = 0$ for $n < 0$, $n \geq n_0$. Set

$c_n = \rho(C_n) =$ rank of C_n

$H_n = n^{th}$ homology of $\left\{C_n \xrightarrow{\partial} C_{n-1}\right\}$

$b_n = \rho(H_n) = n^{th}$ Betti number.

Prove the relation (Euler-Poincaré-Hopf)

$$\Sigma_n (-1)^n c_n = \Sigma_n (-1)^n b_n.$$

(Hint: Use induction on n_0 and the fact that if

$$0 \to G' \to G \to G'' \to 0$$

is an exact sequence of abelian groups, then

$$\rho(G) = \rho(G') + \rho(G'').)$$

If now X is a finite CW complex and

$c_n = \rho(C_n(X)) = \#$ of n-cells in X

$b_n = \rho(H_n(X)) = n^{th}$ Betti number of X

$\chi(X) = \Sigma_n (-1)^n b_n =$ Eüler characteristic of X,

then you have proved

$$\Sigma_n (-1)^n c_n = \chi(X) \quad \text{(formula of Euler-Poincaré).}$$

Exercises on Eüler characteristic and spectral sequences.

22. Let $F \to E \to B$ be a fibration where $\pi_1(B)$ acts trivially

on $H^*(F)$. Suppose that $H^*(B)$, $H^*(F)$ are finitely generated. Prove that

(i) $H^*(E)$ is finitely generated

(ii) $\chi(E) = \chi(B)\chi(F)$ where "χ" means Euler characteristic.

(Hint: Use the Serre spectral sequence to see that $H^*(E)$ can be "no larger" than $H^*(B) \otimes H^*(F)$, and so $H^*(E)$ is finitely generated. To prove (ii) use exercise (21) together with the relation

$$E_{r+1} = \text{homology of } \{E_r, d_r\} .$$

This formula $\chi(E) = \chi(B)\chi(F)$ means that the Euler characteristic is <u>multiplicative</u> for fiber spaces.

23. Let $F \to E \to S^n$ be a fibration over an n-sphere ($n \geq 2$). Deduce the exact sequence (<u>Wang sequence</u>)

$$\cdots \longrightarrow H^p(E) \xrightarrow{i_*} H^p(F) \xrightarrow{\mathcal{D}} H^{p-n+1}(F) \longrightarrow H^{p+1}(E) \xrightarrow{i^*} \cdots$$

where $i: F \rightarrow E$ is the inclusion and where

$$\mathcal{D}(u \cdot v) = \mathcal{D}(u) \cdot v \pm u\mathcal{D}(v).$$

(Hint. Since $H^q(S^n) = 0$ for $q \neq 0, n$ the E_2 term of the spectral sequence looks like

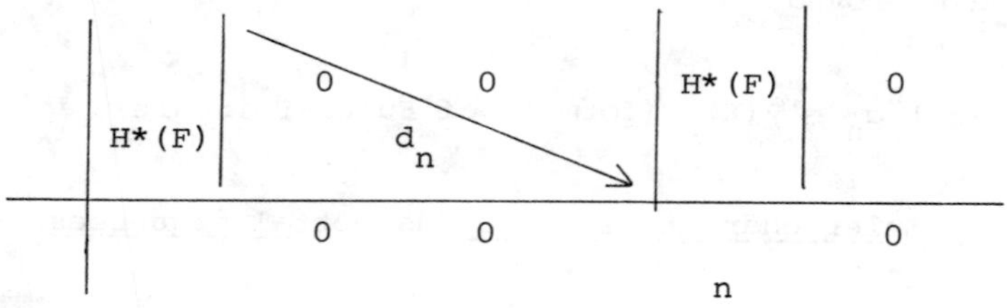

Deduce that the only nonzero differential in the spectral sequence is d_n. From this conclude that

(a) $E_2 = \ldots = E_{n-1}, \; E_{n+1} = E_{n+2} = \ldots = E_\infty$

(b)
$$\begin{array}{ccccccccc} & & H^p(E) & \xrightarrow{i^*} & H^p(F) & \longrightarrow & H^{p-n+1}(F) & \longrightarrow & H^{p+1}(E) & \\ & & \cup & & \wr\Vert & & \wr\Vert & & \cup & \\ 0 & \longrightarrow & E_\infty^{p,0} & \longrightarrow & E_n^{p,0} & \xrightarrow{d_n} & E_n^{p-n+1,n} & \longrightarrow & E_\infty^{p-n+1,n} & \longrightarrow 0 \end{array}$$

The point of this exercise is to illustrate the general principle: When the E_2-term of the spectral sequence has most terms zero, then it simplifies considerably.)

24. Another problem on spectral sequences. Let $S^n \to E \overset{\pi}{\to} B$ be a fibration with fiber a sphere and where $\pi_0(B) = 0$ and $\pi_1(B)$ acts trivially on $H^*(S^n)$. Deduce the exact cohomology sequence (Gysin sequence)

$$\ldots \xrightarrow{\pi^*} H^p(E) \longrightarrow H^{p-n}(B) \xrightarrow{\psi} H^{p+1}(B) \xrightarrow{\pi^*} H^{p+1}(E) \to \ldots$$

where ψ is cup product with a class $e \in H^{n+1}(B)$ (e is the Euler class, defined by obstruction theory in exercise (27) below).

(Hint. The E_2 term in the spectral sequence is

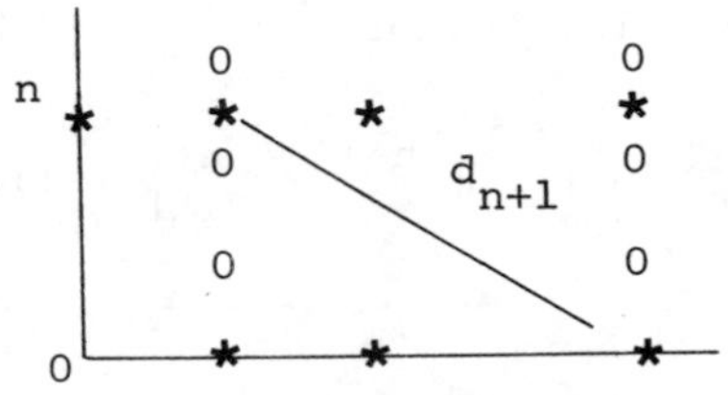

and so $E_2 = \ldots = E_n, \; E_{n+2} = \ldots = E_\infty$ and d_{n+1} is the only nonzero differential. The Euler class $e = d_{n+1}(1)$ where $1 \in H^n(F) \cong E_2^{0,n}$ is the generator of $H^n(S^n)$, and where now the argument proceeds as in the previous exercise using $d_n(\alpha \otimes \beta) = \alpha \otimes d_n(\beta)$ for $\alpha \in H^*(B)$, $\beta \in H^*(F)$).

25. Thom Isomorphism. Let $E \to X$ be an orientable vector bundle of dimension n. Choose a metric on E. Let $D(E) \subset E$ be the unit disk subbundle with respect to this metric, and let $S(E)$ be the unit sphere subbundle. We have a relative fibration $(D(E),S(E)) \to X$; the fibers are pairs (D^n,S^{n-1}). There is a relative version of the Serre spectral sequence where $E_2^{p,q} = H^p(X;H^q(D^n,S^{n-1}))$ and $E_\infty \Rightarrow H^*(D(E),S(E))$. Using this establish the Thom Isomorphism:
There is a class $U \in H^n(D(E),S(E))$ so that
$H^*(X) \cong H^*(D(E)) \xrightarrow{\cup U} H^*(D(E),S(E))$ is an isomorphism.

26. Still another problem on spectral sequences. Let $\mathbf{P}^n \to E \to B$ be a fibration with complex projective space as fiber.

Recall that

$$H^*(\mathbf{P}^n) \cong \mathbf{Z}[x]/(x^{n+1}), \quad x \in H^2(\mathbf{P}^n).$$

Suppose that there exists a class $\zeta \in H^2(E)$ such that $\zeta|\mathbf{P}^n = x$ ($\zeta|\mathbf{P}^n$ means restriction of ζ to fiber). Show that we have the ring isomorphism

$$H^*(E) \cong H^*(B)[\zeta]/(\zeta^{n+1} - \Sigma_{q=1}^{n+1} (-1)^{n+1-q} c_q \zeta^{n+1-q})$$

where $c_q \in H^{2q}(B)$.

(Hint: There is a natural map

$$H^*(B)[\zeta] \longrightarrow H^*(E)$$

coming from $H^*(B) \xrightarrow{\pi^*} H^*(E)$ and the ring structure on $H^*(E)$. By assumption there exists $\zeta_0 \in E_\infty^{0,2}$ corresponding to $x \in E_2^{0,2} \cong H^2(\mathbf{P}^n)$. Thus

$$d_2 x = d_3 x = \ldots = 0.$$

Since $E_2 \cong H^*(B)[x]/(x^{n+1})$, deduce that $E_2 = E_3 = E_4 = \ldots = E_\infty$. Thus

$$H^*(B)[\zeta] \longrightarrow H^*(E)$$

is surjective. <u>Additively</u> there is an isomorphism

$$H^*(E) \cong \oplus_{q=0}^{n} H^*(B)\zeta^q,$$

and so the kernel of the restriction mapping

$$H^*(E) \longrightarrow H^*(\mathbf{P}^n)$$

is $\oplus_{q=1}^{n} H^*(B)\zeta^q$. But $\zeta^{n+1}|\mathbf{P}^n = 0$, and so we obtain a relation in the ring $H^*(E)$ of the form

$$\zeta^{n+1} = \Sigma_{q=1}^{n+1} (-1)^{n-q} c_q \zeta^{n+1-q}, \quad c_q \in H^{2q}(B).$$

Deduce finally that

$$H^*(B)[\zeta]/(\zeta^{n+1} - \Sigma_{q=1}^{n+1}(-1)^{n+1-q} c_q \zeta^{n+1-q}) \longrightarrow H^*(E)$$

is an injection.)

(Note: The classes $c_q \in H^{2q}(B,\mathbf{Z})$ are <u>Chern classes</u> as explained in exercise (30) below. This definition, due to Grothendieck, is the easiest.)

27. <u>A final problem on spectral sequences.</u> Let U_n be the unitary group of $n \times n$ matrices A satisfying $A^{t}\bar{A} = I$. Prove that the cohomology ring

$$H^*(U_n) \cong \wedge(x_1, x_3, \ldots, x_{2n-1})$$

is an exterior algebra with generators in dimensions $1, 3, \ldots, 2n-1$. Thus there is an isomorphism of rings

$$H^*(U_n) \cong H^*(S^1 \times S^3 \times \ldots \times S^{2n-1}).$$

(<u>Hint</u>: Writing $A = (e_1, \ldots, e_n)$ where the e_i are column vectors in $\mathbb{C}^n$, the unitary condition is that $e_1, \ldots, e_n$ should give a unitary frame.

The map

$$\begin{array}{ccc} U_n & \xrightarrow{\pi} & S^{2n-1} \\ \downarrow & & \downarrow \\ A & \xrightarrow{\pi} & e_1 \end{array}$$

gives a fibration

$$\begin{array}{ccc} U_{n-1} & \longrightarrow & U_n \\ & & \downarrow \\ & & S^{2n-1} \end{array}$$

(why?). By induction we have $H^*(U_{n-1}) \cong \wedge(x_1, \ldots, x_{2n-3})$. Now use the Wang sequence (exercise (23)) and show that $\mathcal{D}(x_{2p-1}) = 0$ for $1 \leq p \leq n-1$.)

<u>Exercises on obstruction theory.</u>

In the following exercises on obstruction theory, we assume that $F \to E \to B$ is a fibration where $\pi_0(B) = 0$ and $\pi_1(B)$ acts trivially on the cohomology of the fiber.

28. Euler classes and Euler numbers. Given $F \to E \overset{\pi}{\to} B$, assume that $\pi_i(F) = 0$ for $i < n-1$. Denote by $B^{(k)}$ the k-skeleton of B (always assuming that B is a CW complex). Show that

(a) there exists a cross-section $B^{(n-1)} \overset{\sigma}{\to} E$ (recall that a cross-section σ of $E \to B$ is a map $\sigma: B \to E$ such that $\pi\sigma$ = identity. Think of this as $\sigma(x) \in F_x$ for all $x \in B$).

(b) Any two sections $B^{(n-2)} \underset{\sigma_2}{\overset{\sigma_1}{\longrightarrow}} E$ are homotopic.

(c) The obstruction to finding $B^{(n)} \overset{\sigma}{\to} E$ is given by a cohomology class $c(E) \in H^n(B, H_{n-1}(F))$.

Remark: (a), (b), (c) imply that in a spherical fibration $S^{n-1} \to E \to B$ there exists a unique class $e(E) \in H^n(B,\mathbb{Z})$, the Euler class, such that $S^{n-1} \to E \to B$ has a section over $B^{(n)} \Leftrightarrow e(E) = 0$. In case B is an oriented n-manifold, we may define the Euler number

$$e(E) = \langle e(E), B \rangle .$$

$S^{n-1} \to E \to B$ has a section $\Leftrightarrow e(E) = 0$.

In this exercise use that, for each n-cell e^n on B,

$$E|_{e^n} \sim e^n \times F.$$

29. Let B be a compact, oriented n-manifold and $\mathbb{R}^n \to E_0 \to B$ an oriented vector bundle (give the definition here). Choose a metric in $\mathbb{R}^n \to E_0 \to B$ and let $S^{n-1} \to E \to B$ be the unit sphere-bundle.

Let $e(E)$ be the Euler number as defined in the previous exercise, so that we have a nowhere zero section of $\mathbb{R}^n \to E_0 \to B \Leftrightarrow e(E) = 0$.

(a) Verify that to compute $e(E)$ you do the following:

Find a nonzero section $B^{(n-1)} \xrightarrow{\sigma} E^*$ ($E^* = E$ - zero section) (here we are assuming that B has been triangulated). For each n-cell e_α^n, we have $E|e_\alpha^n \approx e_\alpha^n \times \mathbb{R}^n$ and so σ gives $\partial e_\alpha^n \xrightarrow{\sigma} \mathbb{R}^n - \{0\}$, or equivalently

$$\partial e_\alpha^n \xrightarrow{\sigma_\alpha} S^{n-1}$$

then

$$e(E) = \Sigma_\alpha \text{ degree}(\sigma_\alpha) \quad .$$

Why is it sufficient to use <u>any</u> section σ?

(b) Consider the <u>universal line bundle</u>

$$\mathbb{C} \longrightarrow E \xrightarrow{\pi} \mathbb{P}^1$$

$$\pi^{-1}[z_0,z_1] = \text{line } \mathbb{C}(z_0,z_1) \subset \mathbb{C}^2.$$

Using the Euclidean metric on $\mathbb{C}^2$, the associated sphere bundle is the Hopf fibration

$$S^1 \longrightarrow S^3 \longrightarrow \mathbb{P}^1.$$

Over $\mathbb{P}^1 - \{\infty\}$ ($\infty = [0,1]$) choose the cross-section

$$\sigma([z_0,z_1]) = (1, z_1/z_0).$$

Using this show that the Euler number

$$e(E) = -1.$$

30. <u>Definition of Chern classes using obstruction theory</u>.

(a) Recall the Stiefel manifold $S(n-k+1,n)$ of $n-k+1$

frames in $\mathbb{C}^n$. We computed (cf. exercise 16)

$$\pi_i(S(n-k+1,n) = 0, \quad i < 2k-1$$

$$\pi_{2k-1}(S(n-k+1,n)) \cong \mathbb{Z}.$$

Let $\mathbb{C}^n \to E \to B$ be a complex vector bundle over a CW complex B. Using exercise (28) deduce that there is a class

$$c_k(E) \in H^{2k}(B,\mathbb{Z})$$

such that $\mathbb{C}^n \to E \to B^{(2k)}$ has a field of $n - k + 1$ frames $\Leftrightarrow$ $c_k(E) = 0$. Show that if we have a diagram of mappings

$$\begin{array}{ccc} E & \xrightarrow{F} & E' \\ \downarrow & & \downarrow \\ B & \xrightarrow[f]{} & B' \end{array} \qquad \text{(F linear on fibers)}$$

then $f^*c_k(E') = c_k(E)$. The classes $c_0(E) = 1$, $c_1(E),\ldots,c_n(E)$ are the <u>Chern classes</u> of $\mathbb{C}^n \to E \to B$.

(b) Using exercise (29 (b)), show that for the universal line bundle $\mathbb{C} \to E \to \mathbb{P}^1$

$$c_1(E) = -1 \in H^2(\mathbb{P}^1,\mathbb{Z}).$$

(c) Let $\mathbb{C} \to E_n \to \mathbb{P}^n$ be the universal line bundle over $\mathbb{P}^n$ ($\pi^{-1}(z_0,\ldots,z_n)$ = line $(tz_0,\ldots,tz_n)$ in $\mathbb{C}^{n+1}$). Using 30 (a) above, show that

$$c_1(E_n) = -g$$

where $g \in H^2(\mathbb{P}^n,\mathbb{Z})$ is a generator.

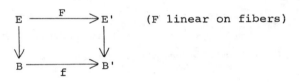

Remark: Given a vector bundle $\mathbb{C}^n \to E \to B$, let $P(E)$ be the bundle of lines in E; thus the fiber $P(E)_x = P(E_x)$ is the set of lines through the origin in $E_x \cong \mathbb{C}^n$. Representing points in $P(E)$ as (x,ξ) (ξ = line in E_x) there is a line bundle $\mathbb{C} \to L(E) \xrightarrow{\pi} P(E)$ where $\pi^{-1}(x,\xi) = \{\text{line } \xi\} \subset E_x$. Consider the fibration $\mathbb{P}^{n-1} \to P(E) \to B$. The first Chern class $c_1(L(E)) \in H^2(P(E),\mathbb{Z})$ restricts to $-\{\text{generator of } H^2(\mathbb{P}^{n-1};\mathbb{Z})\}$ by exercise(30 (a) and (c)).

From this we deduce that $\mathbb{P}^{n-1} \to P(E) \to B$ satisfies the condition in exercise 25 and so, for $\zeta = -c_1(L(E))$,

$$H^*(P(E)) \cong H^*(B)[\xi]/(\zeta^n - \Sigma_{q=1}^n (-1)^{n-q} c_q(E)\zeta^{n-q}).$$

The classes $c_q(E) \in H^{2q}(B)$ are the same Chern classes as those defined by obstruction theory (this is not trivial). The Chern classes are the fundamental invariants for measuring the nontriviality of a vector bundle $\mathbb{C}^n \to E \to B$.

31. Hopf theorem on singularities of a vector field. Given a vector field $\theta = \Sigma_i \theta_i \, \partial/\partial x_i$ in a neighborhood of the origin in $\mathbb{R}^n$ and having an isolated zero at $x = 0$, there is associated an integer, the index $\text{ind}_0(\theta)$ of θ at $x = 0$, defined as follows:

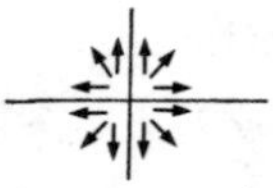

(index +1)

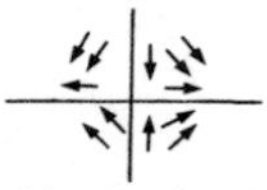

(index -1)

In a small sphere $\|x\| = \varepsilon$, the function $\tilde{\theta}(x) = (\theta_1,\ldots,\theta_n)$ is nonzero, and so $f(x) = \tilde{\theta}(x)/\|\tilde{\theta}(x)\|$ gives a map

$$S^{n-1} \xrightarrow{f} S^{n-1}.$$

Then $\text{ind}_0(\theta) = \text{degree}(f)$.

Let B be a compact, oriented n-manifold and θ a vector field having only isolated zeros.

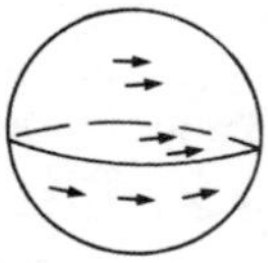

S^2 θ is rotational vector field

We may view θ as a section of the tangent bundle $T \to B$. Prove the formula

$$e(T) = \Sigma_{\theta(x)=0} \operatorname{ind}_x(\theta)$$

<u>Remark</u>: You may assume that B has a CW decomposition with only cells of dimension $\leq n$ (why?), and that θ has zeros only at interior points of the n-cells. Now try to apply exercise(29 (a)).

This proves that $\Sigma_{\theta(x)=0} \operatorname{ind}_x(\theta)$ is <u>independent</u> of the vector field θ. To actually calculate $\Sigma_{\theta(x)=0} \operatorname{ind}_x(\theta)$ for a particular vector field θ, you make a smooth triangulation of B and use the second barycentric subdivision to find a particularly nice vector field. There is a beautiful intuitive discussion of this in Steenrod, Topology of Fiber Bundles, pages 201-203. For this vector field θ, you find the formula

$$\Sigma_{\theta(x)=0} \operatorname{ind}_x(\theta) = \Sigma (-1)^p \{\text{\# of p-cells in cell decomposition of } B\}$$

Putting these formulas together and using exercise (21) gives the final result

$$e(T) = \chi(B) = \Sigma_{\theta(x)=0} \operatorname{ind}_x(\theta)$$

which is a famous theorem of Poincaré-Hopf.

32. Two lemmas.

(a) Let $X' = X \cup_f e^{n+1}$ be obtained by attaching an n+1 cell to X by $\partial e^{n+1} \xrightarrow{f} X$. Using the definition, show that the obstruction to extending $X \xrightarrow{id} X$ to $X' \to X$ is $[f] \in \pi_n(X)$.

(b) Given $A \subset X$, a cohomology class $\alpha \in H^q(X)$, and a cocycle $\tilde{\alpha} \in \tilde{Z}^q(A)$ which represents the restriction $\alpha|A$, show that there is a cocycle $\tilde{\alpha}' \in \tilde{Z}^q(X)$ such that $\tilde{\alpha}'|A = \tilde{\alpha}$. (Here we are using cellular cochains.)

33. Cellular approximation theorem. Suppose that X, Y are CW complexes and $X \xrightarrow{f} Y$ is a continuous map. Show that f is homotopic to a map $X \xrightarrow{g} Y$ which is cellular in the sense that

$$g(X^{(n)}) \subset Y^{(n)}$$

for the respective n-skeleton.

(Hint: For simplicity, assume first that $\dim X < \infty$, and suppose by induction that we have

$$X \xrightarrow{f_n} Y$$

$$f_n \sim f \qquad (\sim \text{ means homotopic to}).$$

$$f_n(X^{(n)}) \subset Y^{(n)}$$

Show now that we have

$$\pi_{n+1}(Y^{(n+1)}, Y^{(n)}) \longrightarrow \pi_{n+1}(Y, Y^{(n)}) \longrightarrow 0$$

and that

$$\pi_{n+1}(Y^{(n+1)}, Y^{(n)})$$

is a free group and the n+1 cells in X (use relative Hurewicz together with $H_q(Y,Y^{(n)}) = 0$ for $0 \leq q \leq n$).

Now let e^{n+1} be an n+1 cell in X. Then $f_n:(e^{n+1},\partial e^{n+1}) \to (Y,Y^{(n)})$ is an element in $\pi_{n+1}(Y,Y^{(n)})$. By the above remark, f_n may be deformed into $\tilde{f}_n: (e^{n+1},\partial e^{n+1}) \to (Y^{(n+1)},Y^{(n)})$. We will be done if we show that, during the homotopy $f_n \sim \tilde{f}_n$, the map may be kept constant on ∂e^{n+1} (why is this necessary?). Thus we must prove the

Lemma: Given $A \subset B \subset Y$, if $(e^{n+1},S^n) \overset{\alpha}{\to} (Y,A)$ is homotopic to $(e^{n+1},S^n) \overset{\beta}{\to} (B,A)$, then α is homotopic to $(e^{n+1},S^n) \overset{\gamma}{\to} (B,A)$ where all maps in the homotopy are constant on $S^n = \partial e^{n+1}$.

Proof of Lemma: Given

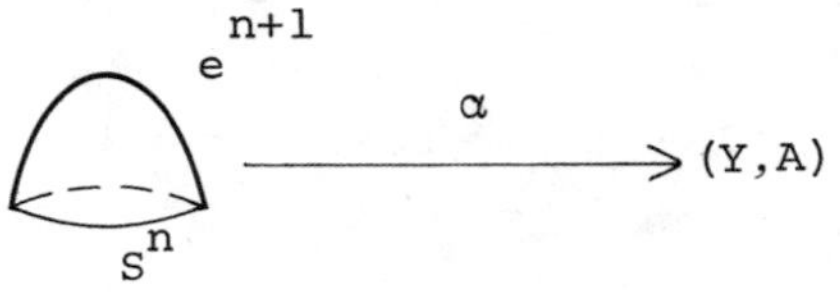

and a homotopy α_t of α with

$$\alpha_0 = \alpha$$

$$\alpha_1 = \beta$$

$$\alpha_t(S^n) \subset A.$$

Put a "collar" around $S^n \hookrightarrow S^n \times I$ and define γ_t by the picture

α_t on the top

α_{st} $(0 \leq s \leq 1)$
on $S^n \times I$ = collar .

Proceeding inductively over the n+1 cells, we now have

$$f_{n+1}: X^{(n+1)} \longrightarrow Y^{(n+1)}$$

$$f_{n+1}|X^{(n)} = f_n$$

$$f_{n+1} \sim f \quad \text{on} \quad X^{(n+1)}$$

Apply the homotopy extension property to extend f_{n+1} to all of X.

What modification is necessary in case dim $X = \infty$?

34. Whitehead products. (Preliminary exercises on wedges and products of spheres.)

(a) Let $S^p \vee S^q \subset S^p \times S^q$ be the usual embedding. Considering the n-sphere as

$$S^n = I^n/\partial I^n,$$

show that

(i) $S^p \times S^q/S^p \vee S^q \sim S^{p+q}$; and

(ii) $H^\ell(S^p \times S^q, S^p \vee S^q; G) \cong H^\ell(S^{p+q}; G)$, $\ell > 0$ for any coefficient group G.

(Hint: For (i), $S^p \times S^q/S^p \vee S^q$

$$= (I^p/\partial I^p \times I^q/\partial I^q)/(I^p/\partial I^p \times *) \cup (* \times I^q/\partial I^q)$$

$$\sim I^{p+q}/\partial I^{q+q} \qquad \text{(why?)}.$$

For (ii) use the previous exercise about excision.)

(Note: This shows that $S^p \times S^q = (S^p \vee S^q) \cup_f e^{p+q}$ where $f\colon S^{p+q-1} \to S^p \vee S^q$, so $[f] \in \pi_{p+q-1}(S^p \vee S^q)$.)
$S^{p+q-1} = \partial e^{p+q}$

(b) Recall the definition of the Whitehead product: Given $S^p \vee S^q \subset S^p \times S^q$, the obstruction to extending the identity map $S^p \vee S^q \xrightarrow{i} S^p \vee S^q$ to $S^p \times S^q \to S^p \vee S^q$ is a cohomology class $\mathcal{O}(i) \in H^{p+q}(S^p \times S^q, S^p \vee S^q; \pi_{p+q-1}(S^p \vee S^q))$ $\cong \pi_{p+q-1}(S^p \vee S^q)$ by (ii) in part (a) above. The Whitehead product

$$w_{p,q} \in \pi_{p+q-1}(S^p \vee S^q)$$

corresponds to $\mathcal{O}(i)$ under this isomorphism.

Using exercise 32(a) and part (i) above, $w_{p,q}$ is the class of the attaching map $\partial e^{p+q} \xrightarrow{f} S^p \vee S^q$ used in constructing $S^p \times S^q \sim (S^p \vee S^q) \cup_f e^{p+q}$.

By considering the cohomology rings of $S^p \times S^q$ and $S^p \vee S^q \vee S^{p+q}$, show that $w_{p,q} \neq 0$. ($w_{p,q}$ is the Whitehead product of the generators $\iota_p \in \pi_p(S^p)$ and $\iota_q \in \pi_q(S^q)$; cf. the remarks below).

If $\alpha \in \pi_p(X)$, $\beta \in \pi_q(X)$ are represented by maps $\tilde{\alpha}\colon S^p \to X$, $\tilde{\beta}\colon S^q \to X$, then the Whitehead product

$$[\alpha,\beta] \in \pi_{p+q-1}(X)$$

is represented by the composite mapping

$$\begin{array}{c} S^{p+q-1} \xrightarrow{\ f\ } S^p \vee S^q \xrightarrow{\ \tilde{\alpha}\vee\tilde{\beta}\ } X \\ \underbrace{\qquad\qquad\qquad\qquad\qquad\qquad}_{[\alpha,\beta]} \end{array}$$

where f is the above attaching map. Thus

$$[\alpha,\beta] = (\alpha \vee \beta)_*(w_{\alpha,q})$$

where $w_{p,q} \in \pi_{p+q-1}(S^p \vee S^q)$ was defined above. Show that the Whitehead product satisfy the relations:

$$[\alpha,\beta] = (-1)^{p,q}[\beta,\alpha] \qquad \text{symmetry}$$

$$[\alpha,[\beta,\gamma]] = [[\alpha,\beta],\gamma] + (-1)^{p,q}[\beta,[\alpha,\gamma]] \quad \text{Jacobi identity,}$$

where $\alpha \in \pi_p(X)$ and $\beta \in \pi_q(X)$.

35. Transgression. (The general case discussed here is not often used. The most important case is the one dealt with in exercise 36.) Let $F \to E \xrightarrow{\pi} B$ be a fibration with $\pi_0(B) = \{0\}$, $\pi_1(B) = \{0\}$, $\pi_1(F) = \{0\}$, so that the Serre spectral sequence is applicable. Consider the cohomology sequences

$$\begin{array}{ccccccc} H^q(E) & \longrightarrow & H^q(F) & \xrightarrow{\ \delta\ } & H^{q+1}(E,F) & \longrightarrow & H^{q+1}(E) \\ & & & & \uparrow{\scriptstyle \pi^*} & & \uparrow{\scriptstyle \pi^*} \\ & & & & H^{q+1}(B,x_0) & \cong & H^{q+1}(B) \end{array} \qquad (q \geq 0).$$

The transgression τ is the map from the subgroup

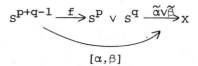

$$T^q(F) = \delta^{-1}\pi^*H^{q+1}(B) \subset H^q(F)$$

to the quotient group

$$G^{q+1}(F) = H^{q+1}(B)/\ker \pi^*$$

defined by

$$\tau(\delta^{-1}\pi^*\alpha) = \alpha \quad .$$

This map is of fundamental importance in the cohomology theory of bundles. Using the proof of the Serre spectral sequence, show that for $q \geq 1$

(i) $T^q(F) \approx E_{q+1}^{0,q} \subset E_2^{0,q} \approx H^q(F)$.

(ii) $G^{q+1}(F) \cong E_{q+1}^{q+1,0}$ is a quotient of $E_2^{q+1,0} \cong H^{q+1}(B)$.

(iii) $\tau = d_{q+1}\colon E_{q+1}^{0,q} \to E_{q+1}^{q+1,0}$.

(Hint: This requires that you look carefully into the construction of the spectral sequence. To get some idea, try the case $q = 1$.

$q = 1$: To prove (i) and (ii) in this case, you must show that $H^2(B) \xrightarrow{\pi^*} H^2(E,F)$ is an isomorphism. This follows from $H_2(E,F) \approx H_2(B)$ (use the relative Hurewicz isomorphism and $\pi_i(E,F) \approx \pi_i(B)$) and the fact that $H^2(B)$, $H^2(E,F)$ are torsion free (why? use $\pi_1(B) = 0 = \pi_1(E,F)$.)

The fact that $\tau = d_2$ in this case follows from the definition of d_2 as a coboundary in an exact cohomology sequence of pairs$(B^{(n)}, B^{(n-2)})$ where $B^{(q)} = \pi^{-1}(X^{(q)})$.

36. <u>Some examples of transgression</u>. (This exercise depends on exercise (35), which you should in any case read over carefully first; cf. also the discussion at the beginning

of Section XI.)

(a) Suppose that

$$\begin{array}{ccc} F & \longrightarrow & E \\ & & \downarrow \pi \\ & & B \end{array}$$

is a fibration and $H^q(E) = 0$ for $q > 0$ (e.g., if E is contractible). Show that

$$\{H^i(F) = 0 \quad \text{for} \quad 0 < i < q\}$$

$$\Leftrightarrow \{H^j(B) = 0 \quad \text{for} \quad 0 < j < q+1\}$$

and that the transgression is an isomorphism

$$\tau: H^q(F) \overset{\approx}{\to} H^{q+1}(B).$$

(<u>Hint</u>: It will suffice to assume that $H^i(F) = 0 = H^j(B)$ for $0 < i < q$, $0 < j < q+1$ and show that

(i) $E_2^{0,q} = \ldots = E_{q+1}^{0,q}$

(ii) $E_2^{q+1,0} = \ldots = E_{q+1}^{q+1,0}$

(iii) $d_{q+1}: E_{q+1}^{0,q} \overset{\approx}{\to} E_{q+1}^{q+1,0}$ is an isomorphism.

The E_2-term of the Serre spectral sequence looks like

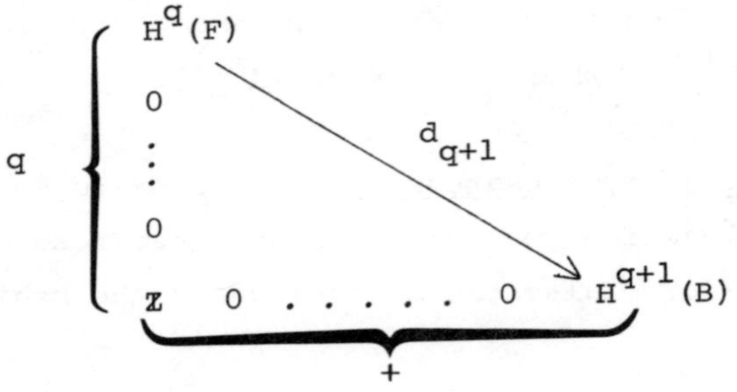

.

From this it follows that $d_r = 0$ on $E_2^{0,q}$ for $2 \leq r \leq q$, and this gives (i) and (ii).

Now then $d_{q+1}: E_{q+1}^{0,q} \to E_{q+1}^{q+1,0}$ must be an isomorphism, since otherwise we would get a nontrivial element in $E_\infty^{0,q}$ or $E_\infty^{q+1,0}$, but $E_\infty = 0$ since $H^*(E) = 0$.)

(b) Let

$$\begin{array}{ccc} K(\pi,n-1) & \longrightarrow & \mathcal{P} \\ & & \downarrow \\ & & K(\pi,n) \end{array}$$

be the path fibration ($n \geq 2$). Show that in the Serre spectral sequence

$$E_2^{0,n-1} = \ldots = E_n^{0,n-1} \cong H^{n-1}(K(\pi,n-1))$$

$$E_2^{n,0} = \ldots = E_n^{n,0} \cong H^n(K(\pi,n-1))$$

$$E_n^{0,n-1} \xrightarrow{d_n} E_n^{n,0} \text{ is an isomorphism.}$$

(Hint: Use part (a) of this exercise above.)

(c) Show that if we have a map of fiber spaces

$$\begin{array}{ccccc} & & E' & \xrightarrow{\tilde{f}} & E \\ & \nearrow & \downarrow & \nearrow & \downarrow \\ F' & & & F & \\ & & \downarrow & & \downarrow \\ & & X' & \xrightarrow{f} & X \end{array}$$

then transgression is natural. (This follows either from the definition or from the spectral sequence interpretation, both of which are given in exercise (35).)

37. <u>Chern classes and transgression</u>. Recall the Stiefel

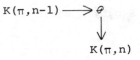

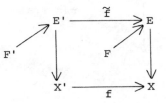

manifold $S(n,N)$ of n-frames in $\mathbb{C}^N$ and that

$$\pi_q(S(n,N)) = 0, \qquad 0 < q < 2N-2n+1 .$$

Using the inclusions $\mathbb{C}^N \subset \mathbb{C}^{N+1}$, we have

$$S(n,N) \subset S(n,N+1)$$

and if we let

$$S(n) = \text{"n-frames in } \mathbb{C}^\infty\text{"}$$

be the infinite Stiefel manifold, then

$$\text{(i)} \qquad \pi_q(S(n)) = 0 \quad \text{for} \quad q > 0$$

(If you don't like "∞" here, take $S(n,N)$ for N arbitrarily large.) Next, recall the Grassmann manifold $G(n)$ of n-planes in $\mathbb{C}^\infty$ and the fibration

$$\begin{array}{ccc} U(n) & \longrightarrow & S(n) \\ & & \big\downarrow \pi \\ & & G(n) \end{array} \qquad \pi(\text{n-frame}) = \begin{array}{c}\text{"n-plane spanned} \\ \text{by the n-frame".}\end{array}$$

Recall that

$$\text{(ii)} \qquad H^*(G(n)) = \mathbb{C}\{x_1, x_2, \ldots, x_n\}$$

is a polynomial algebra with generators

$$x_q \in H^{2q}(G(n)).$$

Finally, recall that for the unitary group, the cohomology

(iii) $$H^*(U(n)) = \wedge(y_1,\ldots,y_n)$$

is an exterior algebra with generators

$$y_q \in H^{2q-1}(U(n)).$$

Prove that each y_q transgresses and

$$\tau(y_q) \equiv x_q \bmod (x_1,\ldots,x_{q-1})$$

(Hint: From equation (i) it follows that $H^q(S(n)) = 0$ for $q > 0$. The E_2 of the spectral sequence looks like

y_3	$\mathbb{Z}$						
y_1y_2	$\mathbb{Z}$						
y_2	$\mathbb{Z}$						
$y_q \in H^{2q-1}(U(n))$	0						
y_1	$\mathbb{Z}$						
	$\mathbb{Z}$	0	$\mathbb{Z}$	0	$\mathbb{Z}\oplus\mathbb{Z}$	0	$\mathbb{Z}\oplus\mathbb{Z}\oplus\mathbb{Z}$
			x_1		$\{x_1^2, x_2\}$		$\{x_1^3, x_1x_2, x_3\}$

$$x_q \in H^{2q}(G(n))$$

Now argue as follows:

$$d_2 y_1 = x_1 \qquad \text{since } H^2(S(n)) = 0$$

$d_2y_2 = 0 = d_3y_3$ because of the 0's in the spectral sequence

$\Rightarrow y_2$ is transgressive and $d_4y_2 \equiv x_2 \bmod (x_1^2)$

$d_2y_3 = 0$ since $d_2y_1y_2x_1 = x_1^2y_2 + x_1x_2y_2 \neq 0$

and $d_3y_3 = d_4y_3 = d_5y_3 = 0$ because of 0's in the spectral sequence

$\Rightarrow y_3$ is transgressive and $d_6y_3 \equiv x_3 \bmod (x_1x_2)$, etc.)

Remark: It can be proved that $x_q \in H^{2q}(G(n))$ is the qth Chern class of the universal vector bundle

$$\begin{array}{ccc} \mathbb{C}^n & \longrightarrow & E \\ & & \downarrow \\ & & G(n) \end{array}$$

Taking into account exercise (30) above, this gives a third definition of Chern classes (these definitions are all the same <u>up to sign</u>!) Roughly speaking, these various definitions have the following advantages:

{Definition 1 using obstruction theory} $\longleftrightarrow$ {useful for "classical" algebraic topology, such as existence of vector fields, etc.}

{Definition 2 using the projective bundle $\mathbb{P}(E)$} {quickest definition and the easiest one to prove the duality theorem, also it is useful in algebraic geometry}

$$\left\{\begin{array}{l}\text{Defn. 3 using}\\ \text{transgression and}\\ \text{cohomology of the}\\ \text{Grassmannians}\end{array}\right\} \quad \left\{\begin{array}{l}\text{this definition is useful in}\\ \text{algebraic geometry, and also}\\ \text{is the definition which shows}\\ \text{how to compute Chern classes}\\ \text{using curvature via deRham's}\\ \text{theorem.}\end{array}\right\}$$

38. <u>(On spectral sequences) Cohomology of</u> $K(\mathbb{Z},3)$. Prove that

$$H^q(K(\mathbb{Z},3)) = \begin{cases} \mathbb{Z} & q = 0,3 \\ \mathbb{Z}/2\mathbb{Z} & q = 6 \\ 0 & \text{otherwise for } 0 \leq q \leq 6 \end{cases} .$$

(Hint: Use the fibration $K(\mathbb{Z},2) \to \mathcal{P}$ together with the
$$\begin{array}{c}\mathcal{P}\\ \downarrow\\ K(\mathbb{Z},3)\end{array}$$
Serre spectral sequence and the fact that

$$H^q(\mathcal{P}) = 0 \quad (q > 0) \quad \text{(why?)}$$

$$\begin{aligned} H^*(K(\mathbb{Z},2)) &= H^*(\mathbb{C}P^\infty) \\ &= \mathbb{Z}[x], \ \ x \in H^2(K(\mathbb{Z},2)). \end{aligned}$$

The E_2 term is

$$\begin{array}{r|cccc} H^*(K(\mathbb{Z},2)) & 0 & & & \\ x^2 & \mathbb{Z} & & & \\ & 0 & & & \\ x & \mathbb{Z} & & & 2xy \\ & 0 & & & 0 \\ & \mathbb{Z} & 0 & 0 & \mathbb{Z} \\ \hline & & & & H^*(K(\mathbb{Z},3)) \\ & & & & y \end{array}$$

where $y \in H^3(K(\mathbb{Z},3))$ is a generator, by Hurewicz.

Then show that $d_2 = 0$ for low values and

$$d_3 x = y \qquad \text{(why?)}$$

$$d_3 x^2 = 2xy \qquad \text{(why?)}$$

$$d_3 y = 0 \qquad \text{(obvious).}$$

$$\Rightarrow E_3^{3,2}/d_3 E_3^{0,4} \cong \mathbb{Z}/2\mathbb{Z} \text{ with generator } xy.$$

But $E_4^{3,2} = 0$ (why? use $E_\infty^{3,2} = 0$).

$$\Rightarrow \ker\{E_3^{3,2} \xrightarrow{d_3} E_3^{6,0}\} = \operatorname{im}\{E_3^{0,4} \xrightarrow{d_3} E_3^{3,2}\}$$

$$\Rightarrow E_3^{6,0} \cong \mathbb{Z}/2\mathbb{Z} \qquad \text{(to prove this you must show that } H^4(K(\mathbb{Z},3)) = 0\text{)}$$

$$\Rightarrow H^6(K(\mathbb{Z},3)) \cong \mathbb{Z}/2\mathbb{Z}.$$

39. Homotopy groups of spheres. Prove that

$$\pi_4(S^2) \cong \mathbb{Z}/2\mathbb{Z}.$$

(Hint: To do this, use the Postnikov tower (= P.T.) for the 2-sphere S^2.

Letting $(S^2)_n$ be the nth stage of the P.T., recall that there is a map

$$S^2 \xrightarrow{f_n} (S^2)_n$$

such that

$$\pi_i(S^2) \cong \pi_i((S^2)_n) \qquad i \leq n$$

$$H_i(S^2) \xrightarrow[(f_n)_*]{\sim} H_i((S^2)_n) \qquad 0 \leq i \leq n+1.$$

<u>Step one</u>: $(S^2)_2 = K(\mathbb{Z},2) = \mathbb{C}P^\infty$ since $\pi_2(S^2) = \mathbb{Z}$.

<u>Step two:</u> There is a fibration

$$\begin{array}{ccc} K(\mathbb{Z},3) & \longrightarrow & (S^2)_3 \\ & & \downarrow \\ & & K(\mathbb{Z},2) = (S^2)_2 \end{array}$$

since $\pi_3(S^2) \cong \mathbb{Z}$. Moreover $H^4((S^2)_3) = 0$. Look in the spectral sequence of this fibering, whose E_2 is, using the previous exercise,

$$\begin{array}{r|ccccc} H^*(K(\mathbb{Z},3)) & & & & & \\ & \mathbb{Z}/2\mathbb{Z} & & & & \\ & 0 & & & & \\ & 0 & & & & \\ H^3(K(\mathbb{Z},3)) \ni y & \mathbb{Z} & & & & \\ & 0 & & & & \\ & 0 & & & & \\ & \mathbb{Z} & 0 & \mathbb{Z} & 0 & \mathbb{Z} \\ \hline & & & x & & x^2 \end{array} \quad H^*(K(\mathbb{Z},2)) \cong \mathbb{Z}[x].$$

Show then that $d_4 y = x^2$, and from this deduce that, for $0 \leq q \leq 6$

$$H^q((S^2)_3) = \begin{cases} \mathbb{Z} & q = 0,2 \\ \mathbb{Z}/2\mathbb{Z} & q = 6 \\ 0 & \text{otherwise} \end{cases}$$

(Note: You must also use that $d_4(yx) = x^3$.)

<u>Step three</u>. Let $\pi = \pi_4(S^2)$. Then the next stage of the P.T. is

$$\begin{array}{ccc} K(\pi,4) & \longrightarrow & (S^2)_4 \\ & & \downarrow \\ & & (S^2)_3 \end{array}$$

and $H^q((S^2)_4) = 0$ for $2 < q \leq 5$. From this deduce that

a) π is a finite abelian group (it might be 0)

b) $H^4(K(\pi,4);\mathbb{Z}) = 0$ and

c) $H^5(K(\pi,4);\mathbb{Z}) \cong \pi$.

Thus the E_2 term of the spectral sequence is

$$\begin{array}{l} H^*(K(\pi,4)) \\ \begin{array}{ccccccc} \pi & & & & & & \\ 0 & & & & & & \\ 0 & & & & & & \\ 0 & & & & & & \\ 0 & & & & & & \\ \mathbb{Z} & 0 & \mathbb{Z} & 0 & 0 & 0 & \mathbb{Z}/2\mathbb{Z} \end{array} \qquad H^*((S^2)_3). \end{array}$$

Deduce from $E_\infty^{s,t} = 0$, $3 \leq s+t \leq 5$, that

$$\begin{array}{cccc} d_6\colon & E_6^{0,5} & \longrightarrow & E_6^{6,0} \\ & \wr\wr & & \wr\wr \\ & E_2^{0,5} & \longrightarrow & E_2^{6,0} \end{array}$$

is an isomorphism. Thus, $\pi \cong \mathbb{Z}/2\mathbb{Z}$.

40. <u>Classification theorem for complex vector bundles</u>. Let X be a CW complex of dimension n and $\mathrm{Vect}^r(X)$ be the equivalence classes of r-plane bundles $\mathbb{C}^r \to E \to X$. Denote by

$G(r,N)$ the Grassmanian of r-planes in $\mathbb{C}^N$ and by

$$\begin{array}{ccc} \mathbb{C}^r & \longrightarrow & U_r \\ & & \downarrow \pi \\ & & G(r,N) \end{array}$$

the <u>universal vector bundle</u> (thus π^{-1}(r-plane) = the same r-plane considered as a vector space). Show that the assignment

$$\begin{array}{ccc} [X,G(r,N)] & \longrightarrow & \mathrm{Vect}^r(X) \\ \downarrow & & \\ f & \longrightarrow & f^*U_r \end{array}$$

is a bijection for $2N - 2r + 1 > n$.

<u>Proof</u>: Recall the fibration $U_r \to S(r,N) \to G(r,N)$, where $S(r,N)$ is the Stiefel manifold of r-frames in $\mathbb{C}^N$, and also that

$$\pi_i(S(r,N)) = 0 \quad \text{for} \quad i < 2N-2r+1$$

$$\Rightarrow \pi_i(S(r,N)) = 0 \quad \text{for} \quad i < n+1$$

under the conditions of the theorem. Let's first show that every vector bundle $\mathbb{C}^r \to E \to X$ is of the form f^*U_r for some map $X \to G(r,N)$. Letting $X^{(k)}$ be the k-skeleton, suppose we have

$$X^{(k)} \xrightarrow{f^{(k)}} G(r,N)$$

$$f^{(k)*}U_r \quad \cong \quad E|X^{(k)}.$$

Now the restriction of E to each $k+1$ cell e_{k+1} is trivial (why?), so that we have

$$E|e_{k+1} \cong e_{k+1} \times \mathbb{C}^r$$

$$\partial e_{k+1} \xrightarrow{f^{(k)}} G(r,N)$$

$$(*) \qquad \Rightarrow f^{(k)*}E|\partial e_{k+1} \cong \partial e_{k+1} \times \mathbb{C}^r.$$

This gives us a lifting of $\partial e_{k+1} \xrightarrow{f^{(k)}} G(r,N)$ to

$$\begin{array}{ccc} & \overset{g}{\nearrow} & S(r,N) \\ & & \downarrow \\ \partial e_{k+1} & \xrightarrow[f^{(k)}]{} & G(r,N) \end{array}$$

by taking the r-frame to be the coordinate frame in $\mathbb{C}^r$ under the isomorphism (*).

Since $\pi_k(S(r,N)) = 0$, g extends to e_{k+1}, and in this way we may extend $f^{(k)}$ over the (k+1)-skeleton (why?). Thus

$$[X,G(r,N)] \longrightarrow \text{Vect}^r(X)$$

is onto, and now a relative version of the same argument shows that it is one-to-one.)

<u>Remark</u>: A special case of this theorem is the isomorphism

$$\text{Vect}^r(S^n) \cong \pi_n(G(r,N)) \cong \pi_n(BU_r)$$

(N large relative to r,n)

Suppose we call two vector bundles E, E' on a space X <u>stably isomorphic</u> if

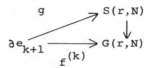

$$E \oplus t^{\ell} \approx E' \oplus t^{\ell'}$$

where t^{ℓ} is the trivial bundle of rank ℓ. The equivalence classes of stable vector bundles will be denoted by K(X), and then the above theorem gives

$$K(X) \cong [X, BU].$$

Returning to the n-sphere, we have then

$$K(S^n) \cong \pi_n(BU),$$

and so Bott periodicity computes the stable vector bundles over spheres.

41. Axioms for homology. Suppose we are given a covariant functor

$$X \longrightarrow \mathcal{H}_*(X) = \oplus_{p \geq 0} \mathcal{H}_p(X)$$

from finite CW complexes to abelian groups such that

(i) $X \xrightarrow{f} Y$ induces $\mathcal{H}_*(X) \xrightarrow{f_*} \mathcal{H}_*(Y)$ which depends only on the homotopy class of f.

(ii) $\mathcal{H}_*(pt) = \begin{cases} \mathbb{Z} & * = 0 \\ 0 & * > 0 \end{cases}$

(iii) given $X \subset Y$, if we define

$$\mathcal{H}_*(Y,X) \cong \mathcal{H}_*(Y/X)$$

(this property forces excision to be true), then the exact homology sequence for a pair holds.

Show that $\mathcal{H}_*(X) = H_*(X;\mathbb{Z})$ is ordinary homology.

(Hint: By (ii), this is true if dim $X = 0$. Suppose by induction it is true when dim $X < n$. If D^n is the n-disc with $\partial D^n = S^{n-1}$, then $\mathcal{H}_*(D^n) = \{\begin{smallmatrix}\mathbb{Z}, & *=0 \\ 0, & n>0\end{smallmatrix}\}$ by (i) and $_*(S^{n-1}) \cong \{\begin{smallmatrix}\mathbb{Z}, & *=0,n-1 \\ 0, & \text{otherwise}\end{smallmatrix}\}$ by induction. The exact homology sequence then gives

$$(*) \qquad \mathcal{H}_*(D^n, \partial D^n) \cong H_*(D^n, \partial D^n).$$

Suppose now that $X = Y \cup_f e^n$ where dim $Y \leq n-1$. Then we have

$$\begin{array}{ccccccc} H_p(Y) & \longrightarrow & H_p(X) & \longrightarrow & H_p(X,Y) & \xrightarrow{\partial} & H_{p-1}(Y) \\ & & & & \wr\wr & & \\ \mathcal{H}_p(Y) & \longrightarrow & \mathcal{H}_p(X) & \longrightarrow & \mathcal{H}_p(X,Y) & \longrightarrow & \mathcal{H}_{p-1}(Y) \end{array}$$

where the isomorphism $\mathcal{H}_p(X,Y) \approx H_p(X,Y)$ follows by (*) above.

Taking $p = n$, we get a map $\mathcal{H}_n(X) \to H_n(X)$ which is an $\cong$. For $p = n-1$, we have

$$H_{n-1}(X) \cong H_{n-1}(Y)/H_n(X,Y)$$

and

$$\mathcal{H}_{n-1}(X) \cong \mathcal{H}_{n-1}(Y)/\mathcal{H}_n(X,Y).$$

This gives a map $\mathcal{H}_{n-1}(X) \to H_{n-1}(X)$ which is an isomorphism. The remaining $\mathcal{H}_p$, H_p for $p < n-1$ are already isomorphic (why?).

Now complete the proof by induction on the number of n-cells in X.)

Remark: The analogous theorem for cohomology is also true with the same proof. In general, a cohomology theory is a contravariant functor

$$X \longrightarrow K^*(X)$$

from spaces X to abelian groups which satisfies the cohomology acioms corresponding to (i) and (ii). Using Bott periodicity, and denoting by S^nX the n-fold suspension of X, one obtains K-theory with

$$K^n(X) \underset{\text{defn.}}{\equiv} K(S^nX)$$

as an extraordinary cohomology theory.

42. Little problems on filtrations.

(i) Show that if $A = A_0 \supset A_1 \supset \ldots \supset A_m = \{0\}$ and $B = B_0 \supset B_1 \supset \ldots \supset B_n = \{0\}$ are filtered abelian groups, and if $f: A \to B$ is a filtration preserving homomorphism inducing isomorphisms $A_p/A_{p+1} \xrightarrow{\sim} B_p/B_{p+1}$, then f itself is an isomorphism.

(ii) Using (i) show that if a map of Serre spectral sequences induces an isomorphism on E_∞, then the abutments are isomorphic.

(iii) As an example of the failure of the converse of (ii), let X be a CW complex with base point $x_0 \in X$. Consider the trivial fibrations $X \to x_0$ and $X \xrightarrow{id} X$. The identity map $X \to X$ induces a map of these fibrations which is an isomorphism on the abutment of the spectral sequences (which is $H^*(X)$ in both cases). Show from the definitions that this map is not an isomorphism on E_∞ unless $H^*(X) = 0$ for $* > 0$.

43. The comparison theorem for spectral sequences. Let $E^r_{p,q}$, $E'^r_{p,q}$ be two homology spectral sequences (same as cohomology spectral sequences with upper and lower indices interchanged, and with arrows reversed; also the proof of the Serre spectral sequence for homology is the same as for cohomology). Suppose

that we have a map

$$E^r_{p,q} \longrightarrow E'^r_{p,q} \quad \text{(for all } r\text{, and the map commutes with } d_r\text{'s)}$$

such that

$$E^2_{0,q} \cong E'^2_{0,q}$$

and

$$E^\infty_{p,q} \cong E'^\infty_{p,q} \quad \text{for all } p,q \geq 0.$$

Show that $E^2_{p,0} \cong E'^2_{p,0}$ for all $p \geq 0$.

<u>Remark</u>: This may be interpreted as saying that if we have a map between fiber spaces

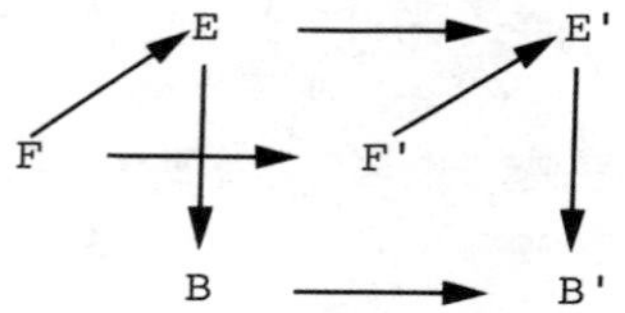

inducing isomorphisms on $H_*(F) \xrightarrow{\sim} H_*(F')$ and $H_*(E) \xrightarrow{\sim} H_*(E')$, then $H_*(B) \to H_*(B')$ is also an isomorphism (compare the previous exercise).

<u>Step one</u>: Show that

$$E^2_{p,0} \cong E'^2_{p,0} \text{ for a } \underline{\text{fixed}}\ p \Rightarrow$$

$$E^2_{p,q} \cong E'^2_{p,q} \text{ for all } q.$$

<u>Step two:</u> Prove that $E^2_{0,0} \cong E'^2_{0,0}$ and

$$E^2_{1,0} \cong E'^2_{1,0}.$$

<u>Step three</u>: Check that $E^3_{0,1} \overset{\sim}{\to} E'^3_{0,1}$ (use $E^\infty_{0,1}$);

$$E^2_{0,1} \overset{\sim}{\to} E'^2_{0,1} \text{ and } E^3_{2,0} \overset{\sim}{\to} E'^3_{2,0} \quad (\text{use } E^\infty_{2,0}).$$

Use the exact sequence

$$0 \to E^3_{2,0} \to E^2_{2,0} \to \operatorname{Ker}\{E^2_{0,1} \to E^3_{0,1}\} \to 0$$

to conclude that

$$E^2_{2,0} \overset{\sim}{\to} E'^2_{2,0}.$$

<u>Step four</u>: For general p, we attack $E^r_{p,0}$ by descending induction on $r \geq 2$ until we find $E^2_{p,0} \overset{\sim}{\to} E'^2_{p,0}$.

First $E^{p+1}_{p,0} \overset{\sim}{\to} E'^{p+1}_{p,0}$ (why? use $E^\infty_{p,0}$) and $E^{p+1}_{0,p-1} \overset{\sim}{\to} E'^{p+1}_{0,p-1}$ (why?). Using $E^2_{0,p-1} \overset{\sim}{\to} E'^2_{0,p-1}$ and ascending induction on r, show that $E^p_{0,p-1} \overset{\sim}{\to} E'^p_{0,p-1}$. Conclude that $E^p_{p,0} \overset{\sim}{\to} E'^p_{p,0}$.

<u>Step five</u>: Using ascending induction on k, prove that

$$E^k_{p-r,s} \overset{\sim}{\to} E^k_{p-r,s} \quad \text{for } 2 \leq k \leq r,\ s \geq 0.$$

In particular

$$E^r_{p-r,r-1} \overset{\sim}{\to} E'^r_{p-r,r-1}.$$

Using this and descending induction on $k \geq 1$, prove that $E^{r+k}_{p-r,r-1} \overset{\sim}{\to} E'^{r+k}_{p-r,r-1}$. In particular,

$$E^{r+1}_{p-r,r-1} \overset{\sim}{\to} E'^{r+1}_{p-r,r-1}.$$

<u>Step six</u>: Show that image $\{E^r_{p,0} \to E^r_{p-1,r-1}\} \overset{\sim}{\to}$ image $\{E'^r_{p,0} \to E'^r_{p-1,r-1}\}$ using

$$E^{r+1}_{p-r,r-1} \overset{\sim}{\to} E'^{\,r+1}_{p-r,r-1} \quad \text{and} \quad \ker\{E^{r}_{p-r,r-1} \longrightarrow E^{r}_{p-2r,2r-2}\}$$

$$\cong \ker\{E'^{\,r}_{p-r,r-1} \longrightarrow E'^{\,r}_{p-2r,2r-2}\}.$$

Conclude that $E^{r}_{p,0} \overset{\sim}{\to} E'^{\,r}_{p,0}$ using $E^{r+1}_{p,0} \overset{\sim}{\to} E'^{\,r+1}_{p,0}$.

44. <u>Homotopy of wedges of 2-spheres</u>. Let $X = S^2_1 \vee \ldots \vee S^2_m$ be a wedge of 2-spheres. Then the second homotopy group

$$\pi_2(X) \cong \oplus^{m}_{i=1} \pi_2(S^2_i) \cong \mathbb{Z}^m,$$

and the <u>Whitehead products</u>

$$\pi_{ij} = [S^2_i, S^2_j] \in \pi_3(X).$$

Show that these products for $i \leq j$ give a free $\mathbb{Z}$-basis for $\pi_3(X)$. Thus

$$\pi_3(\vee^{m}_{i=1} S^2_i) \cong \mathbb{Z}^{m(m+1)/2}.$$

(Hint: This problem is a good exercise in Postnikov towers. Since $\pi_2(X) \cong \oplus^{m}_{i=1} \pi_2(S^2_i)$, the P.T. for X begins with

$$X_2 = \underset{(m \text{ factors})}{\Pi} K(\mathbb{Z},2).$$

The next step is

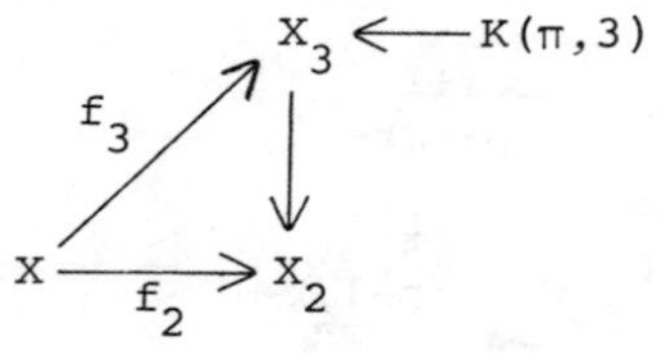

$(\pi = \pi_3(X))$

We want to determine what π is. From the exact cohomology sequence of (X_2,X) we have

$$\begin{array}{ccccccc} H^3(X,\pi) & \longrightarrow & H^4(X_2,X;\pi) & \longrightarrow & H^4(X_2,\pi) & \longrightarrow & H^4(X,\pi) \\ \| & & & & & & \| \\ 0 & & & & & & 0 \end{array}$$

Since $\pi_i(X_2,X) = 0$ for $0 \le i \le 3$ (why?), the pair (X_2,X) begins with cells in dimension 4, and so

$$H^4(X_2,X;\pi) \cong \operatorname{Hom}(H_3(X_2,X),\pi)$$
$$\cong \operatorname{Hom}(\pi,\pi) \quad \text{(why?)}.$$

Thus we have

$$0 \longrightarrow \operatorname{Hom}(\pi,\pi) \longrightarrow H^4(X_2,\pi) \longrightarrow 0.$$

Since $H_*(X_2,\mathbb{Z})$ is free, we obtain then

$$0 \longrightarrow \operatorname{Hom}(\pi,\pi) \xrightarrow{\sim} \operatorname{Hom}(H_4(X_2),\pi) \longrightarrow 0.$$

From this we see that π is free and

$$\operatorname{Hom}(\pi;\mathbb{Z}) \cong H^4(X_2;\mathbb{Z}) \cong \mathbb{Z}^{m(m+1)/2}.$$

It remains to identify $\pi \cong \mathbb{Z}^{m(m+1)/2}$ as having a basis given by the Whitehead products $[S_i^2,S_j^2]$ $(i \le j)$. To do this, replace $K(\mathbb{Z},2)$ by $\mathbb{C}P^2$ in the P.T. for X. This is O.K. since $\pi_i(\mathbb{C}P^2) \cong \pi_i(K(\mathbb{Z},2))$ for $0 \le i \le 4$. Thus we have

$$\begin{array}{ccc} & & X_3 \longleftarrow K(\pi,3) \\ & & \downarrow \\ X & \xrightarrow{f_2} & \Pi\mathbb{C}P^2 \\ \cup & & \cup \\ \vee\mathbf{P}^1 & & \vee\mathbf{P}^1 \end{array}$$

($\vee\mathbf{P}^1$ is the 2-skeleton of $\Pi\mathbb{C}P^2$).

The k-invariant is by definition the primary obstruction to finding a section to the inclusion

$$\vee\mathbf{P}^1 \hookrightarrow \Pi\mathbb{C}P^2.$$

The primary obstruction is a class in

$$H^4(\Pi\mathbb{C}P^2, \pi_3(\vee\mathbf{P}^1)).$$

Recall that the Whitehead product of the generators of $\pi_2(S_0^2), \pi_2(S_1^2)$ is the obstruction to extending the identity $S_0^2 \vee S_1^2 \to S_0^2 \vee S_1^2$ to a map

$$\begin{array}{ccc} S_0^2 \times S_1^2 & \overset{f}{\searrow} & \\ \cup & & \\ S_0^2 \vee S_1^2 & = & S_0^2 \vee S_1^2. \end{array}$$

This essentially completes the desired identification, once we show that the usual Hopf map $S^3 \to S^2$ represents the Whitehead product of the generator of $\pi_2(S^2)$ with itself (Why? Consider the 4-skeleton of $\Pi\mathbb{C}P^2$: this is $(\vee\mathbb{C}P^2) \vee (\Pi\mathbb{C}P^1)$.)

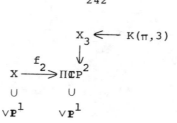

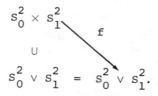

PROGRESS IN MATHEMATICS

Already published

PM 1 Quadratic Forms in Infinite-Dimensional Vector Spaces
Herbert Gross
ISBN 3-7643-1111-8, 431 pages, $22.00, paperback

PM 2 Singularités des systèmes différentiels de Gauss-Manin
Frédéric Pham
ISBN 3-7643-3002-3, 339 pages, $18.00, paperback

PM 3 Vector Bundles on Complex Projective Spaces
C. Okonek, M. Schneider, H. Spindler
ISBN 3-7643-3000-7, 389 pages, $20.00, paperback

PM4 Complex Approximation, Proceedings, Quebec, Canada, July 3-8, 1978
Edited by Bernard Aupetit
ISBN 3-7643-3004-X, 128 pages, $10.00, paperback

PM 5 The Radon Transform
Sigurdur Helgason
ISBN 3-7643-3006-6, 202 pages, $14.00, paperback

PM 6 The Weil Representation, Maslov Index and Theta Series
Gérard Lion, Michèle Vergne
ISBN 3-7643-3007-4, 345 pages, $18.00, paperback

PM 7 Vector Bundles and Differential Equations
Proceedings, Nice, France, June 12-17, 1979
Edited by André Hirschowitz
ISBN 3-7643-3022-8, 255 pages, $14.00, paperback

PM 8 Dynamical Systems, C.I.M.E. Lectures, Bressanone, Italy, June 1978
John Guckenheimer, Jürgen Moser, Sheldon E. Newhouse
ISBN 3-7643-3024-4, 300 pages, $14.00, paperback

PM 9 Linear Algebraic Groups
T. A. Springer
ISBN 3-7643-3029-5, 304 pages, $18.00, hardcover

PM10 Ergodic Theory and Dynamical Systems I
A. Katok
ISBN 3-7643-3036-8, 352 pages, $20.00, hardcover

PM11 18th Scandinavian Congress of Mathematicians, Aarhus, Denmark, 1980
Edited by Erik Balslev
ISBN 3-7643-3034-6, 528 pages, $26.00, hardcover

PM12 Séminaire de Théorie des Nombres, Paris 1979-80
Edited by Marie-José Bertin
ISBN 3-7643-3035-X, 408 pages, $22.00, hardcover

PM13 Topics in Harmonic Analysis on Homogeneous Spaces
Sigurdur Helgason
ISBN 3-7643-3051-1, 142 pages, $12.00, hardcover

PM15 Representations of Real Reductive Lie Groups
David A. Vogan, Jr.
ISBN 3-7643-3037-6, 771 pages, $35.00, hardcover

PROGRESS IN PHYSICS

Already published

PPh1 Iterated Maps on the Interval as Dynamical Systems
Pierre Collet and Jean-Pierre Eckmann
ISBN 3-7643-3026-0, 256 pages, $16.00 hardcover

PPh2 Vortices and Monopoles, Structure of Static Gauge Theories
Arthur Jaffe and Clifford Taubes
ISBN 3-7643-3025-2, 275 pages, $16.00 hardcover

PPh3 Mathematics and Physics
Yu. I. Manin
ISBN 3-7643-3027-9, 111 pages, $10.00 hardcover